"Through its stunning clarity and intellectual richness, this book makes a major contribution to digital humanities. The author combines historical and practical material to introduce the principles of computational thinking. Vivid examples and step-by-step instruction provide excellent demonstrations of these principles. The discussion of mathematics, logic, and statistics is succinct and informative and the demonstration of applications is easy to follow. The organization and language of the book support independent learning for those interested in acquiring the fundamentals of computational literacy for humanities research. But even for those who may not pursue future applications of its principles, this book should be read for the insights it provides into the computational functions that underpin many aspects of contemporary life. I wish I had had this book while I was still teaching, but I would recommend it without reservation to anyone interested in or involved with computation in the humanities."

*Johanna **Drucker**, Distinguished Professor and Breslauer*
Professor Emerita, Department of Information Studies,
University of California, Los Angeles, USA

W0018672

Computational Literacy for the Humanities

Computational Literacy for the Humanities provides an introduction to mathematics and programming that is specifically designed for use by those engaged in the humanities. Linking mathematical concepts and computational skills, the chapters in this book explore humanistic questions from diverse fields, such as art, history, and literature.

The book helps to advance computational and digital literacy by showing that each mathematical concept has a history, and each technique has a meaning. Rather than viewing mathematics and computer programming as purely instrumental, they are integrated into the process of achieving greater understanding of humanistic phenomena. Algorithms, data, statistics and networks are taught critically within the book, whilst the authors also make a concerted effort to expose the internal biases of these tools. They also demonstrate the applicability of quantification and computation for the promotion of diversification and inclusivity within the humanities. All exercises are designed as an opportunity to gain hands-on mathematical and computational experience, whilst critically exploring and interpreting humanistic phenomena.

Computational Literacy for the Humanities shows readers how to engage with data in a way that is challenging, yet meaningful and empowering. It will be of interest to scholars and students working across the humanities and should be of particular interest to those working in digital humanities.

Zef M. Segal is a senior lecturer at the School of Media Studies at the College of Management Academic Studies. His research centers on 19th-century mobility, space, and communication, with a particular focus on the historical analysis of journals and maps. A strong advocate for the digital humanities and social sciences in Israel, he spearheaded the development of the Digital Humanities track at the Open University and was among the founding faculty members of the first Israeli bachelor's degree in Digital Humanities at the University of Haifa. His most recent publications include *Motion in Maps, Maps in Motion* (2020), and *The Political Fragmentation of Germany* (2019). His forthcoming book, *Culture in the Age of the Enlightenment: Reshaping the Private Sphere*, will be published in Hebrew by the Open University of Israel.

Nurit Melnik is a senior lecturer in linguistics at the Open University of Israel. Her studies focus on diverse topics: constructions and the relationship between form and meaning, agreement variations in Modern Hebrew, lexical and syntactic flexibility and "frozen" expressions, grammar engineering and the cognitive aspects behind language change. Her work is situated at the intersection of theoretical linguistics, computational linguistics, and corpus linguistics.

Computational Literacy for the Humanities

Mathematics and Programming in Context

Zef M. Segal and Nurit Melnik

Routledge
Taylor & Francis Group

LONDON AND NEW YORK

Designed cover image: Yael Segal

First published 2025
by Routledge
4 Park Square, Milton Park, Abingdon, Oxon OX14 4RN

and by Routledge
605 Third Avenue, New York, NY 10158

Routledge is an imprint of the Taylor & Francis Group, an informa business

British Library Cataloguing-in-Publication Data
A catalogue record for this book is available from the British Library

ISBN: 9781032820729 (hbk)
ISBN: 9781032788814 (pbk)
ISBN: 9781003502814 (ebk)

DOI: 10.4324/9781003502814

Typeset in Times New Roman
by Newgen Publishing UK

Contents

Figures

Preface

In mid-2018, we began conceptualizing a unique course as part of a new undergraduate program in Digital Humanities at the Open University of Israel. Our intention was to equip students with foundational mathematical knowledge and programming skills – not merely as technical applications, computational formulas, or abstract theories, but as an integral part of humanities studies. The course, titled "Who's Afraid of Numbers?" launched in February 2022 and continues to this day, attracting a diverse group of students. Some participants have backgrounds in computer science or mathematics and wish to explore their application in the humanities, while others are newcomers eager to gain computational literacy.

This book seeks to capture the interactive spirit of our course, maintaining its hands-on approach, which encourages students to engage in small research projects while learning mathematical and computational concepts. These projects highlight both the benefits and limitations of computational methods, demonstrating their practical efficiency alongside the critical thinking required for meaningful interpretation. This emphasis underscores the importance of close reading across all areas of humanities.

We extend our gratitude to everyone who made the course, now transformed into a book, possible. Special thanks go to Prof. Ophir Münz-Manor for championing the establishment of the Digital Humanities program at the Open University, and to the Digital Humanities and Social Sciences Hub and its team for their unwavering support. We also wish to acknowledge Prof. Judith Gal-Ezer, Prof. Reut Tsarfaty, and Dr. Sinai Rusinek for their valuable early-stage suggestions, and Dr. Orly Stettiner for her meticulous feedback on the style and pedagogy of the course.

Introduction

Computational Literacy for the Humanities: Mathematics and Programming in Context is a groundbreaking textbook created to bridge the gap between the humanities and the increasingly digital world we live in. As technology permeates every aspect of daily life, computational literacy has become an essential skill – not just for scientists and engineers, but also for those who study human culture, history, and society. This book is here to empower you, particularly if you come from a humanities background, with the mathematical and computational tools you need to navigate, interpret, and contribute to the digital age. We assume no background in mathematics or computer programming, apart from an inherent curiosity.

I.1 Aims and approach

At its core, this book addresses a fundamental challenge: the perceived divide between the humanities and STEM (Science, Technology, Engineering, and Mathematics) fields. Traditionally, mathematics and computer science have been viewed as separate from the study of art, literature, history, and other humanistic disciplines. However, in today's digital era, this separation isn't just unnecessary; it can be counterproductive. The digital turn has transformed how we analyze texts, visualize historical data, and understand social networks. Engaging with mathematical concepts and computational techniques isn't simply about using tools – it's about integrating them into your research and understanding of the world.

Computational literacy, which is the explicit aim of this book, is the ability to effectively understand, use, and apply computing and mathematical concepts and tools, along with the awareness of the limitations of these techniques (Vee 2013; Selisker 2016; Jacob and Warschauer 2018; Armand and Henriot 2023). Unlike expertise, literacy is not the ability to master every complex computational process or write advanced algorithms. Instead, computational literacy "empowers people by giving them tools to think for themselves, to ask intelligent questions of experts, and to confront authority confidently" (Steen 2001, 2, originally about quantitative literacy).

The most recent revolution in computer technology is the introduction of generative AI, which seemingly reduces the need for mathematical and computational skills. However, while generative AI offers impressive capabilities, it remains a

DOI: 10.4324/9781003502814-1

tool that requires knowledgeable, responsible, and creative users to maximize its potential effectively. Computational literacy ensures that people not only use AI but engage with it in meaningful, informed, and impactful ways, shaping the future of technology rather than being passive recipients of its outcomes.

In *Computational Literacy for the Humanities*, you'll find a novel approach that explores the interconnectedness of mathematics, computation, and humanistic inquiry.

Mathematics serves as the foundation, but not as abstract theory. You'll explore essential mathematical concepts like logic, set theory, statistics, and network theory, presented in ways that directly apply to humanistic inquiry. Each concept is contextualized as a tool that has evolved with human culture, making it accessible and relevant to your studies. The intent is not to cover all the math that one might encounter, but rather to introduce key concepts that serve as useful entry points to the computational study of humanistic materials.

Computation builds on this mathematical groundwork, introducing you to programming with a focus on Python, a powerful language commonly used across disciplines. Here, you'll learn to think algorithmically, structure data, and develop programs to tackle complex problems in the humanities. This introduction will provide you with programming fundamentals, and possibly motivate you to further engage with coding.

Humanistic Interpretation ties everything together. Throughout the book, case studies from diverse fields such as art, history, and literature demonstrate the advantages of using mathematical and computational tools. Whether you're analyzing literary texts, visualizing historical timelines, or mapping social networks, the emphasis is on using these tools to gain new insights into human culture and history.

This holistic approach views mathematics and computation not just as means to an end; they're deeply woven into humanistic interpretation. Mathematical and computational concepts are presented with a brief historical introduction which provides cultural contextualization. Furthermore, these concepts are explained and illustrated with data from various humanistic domains. All exercises are designed as an opportunity to gain hands-on mathematical and computational experience while critically exploring and interpreting humanistic phenomena.

We do not aim to create new mathematicians or programmers or to replace humanistic approaches with automated or quantitative methods. Instead, we hope this book will inspire readers to see new possibilities in their fields of interest.

1.2 Overview of the book

The book is divided into five parts: (I) basic logical concepts, (II) programming fundamentals, (III) reading and creating charts, (IV) statistics for the humanities, and (V) an introduction to network theory.

Part I: Basic logical concepts

Part I consists of one chapter, "Sets, Numbers, and Functions", which introduces key mathematical ideas that are both accessible and practical for humanistic study. It explores the world of numbers, covers the basics of set theory, and presents a formal definition of functions. This broader perspective on functions expands the concept you might remember from school, now applied to sets that go beyond numbers, reflecting the types of materials you'll often work with in the humanities, such as words, texts, and cultural artifacts.

Part II: Programming fundamentals

Part II consists of two chapters dedicated solely to computational thinking and programming. Chapter 2, "Algorithmics", is an introductory chapter on algorithmic thinking. It focuses on problem solving, especially those in the humanities that can be addressed computationally. You'll learn about pseudocode and flowcharts as tools to express solutions for search algorithms in lists and graphs, with a focus on abstraction and modularization. Chapter 3, "Introduction to the Python Programming Language", dives into the fundamentals of programming in general, and Python programming in particular. We start with variables, values, and simple operations, and then move on to conditions and loops, supporting your first implementations of simple algorithms. The final section introduces file input–output operations, allowing you to begin analyzing literary texts with two short stories, "The Haunted Mind" by Nathaniel Hawthorne and "Two Kinds" by Amy Tan.

Part III: Reading and creating charts

Part III shifts the focus to graphical representation of data, helping you develop the skills to critically read and create charts. Diagrams appear in a variety of forms and ways that combine computational, verbal and aesthetic elements. Chapter 4, "Reading charts", guides you in interpreting and analyzing a variety of charts. It introduces scatter plots, line charts, pie charts, histograms and network diagrams, unpacking the assumptions behind each type. Throughout the chapter we will explore visualizations used over the last 200 years, including diagrams created by Florence Nightingale, John Herschel, and William Playfair, as well as more recent examples published in the 21st-century newspapers and journals.

Chapter 5, "Creating Charts", builds on the previous chapter and guides you in generating and building your own charts using spreadsheets. Through case studies, you will create pie charts and scatter plots to explore 19th-century art history, timelines to study political stability in parliamentary democracies, and histograms to analyze literature. In the literary analysis we will continue the investigation of *The Haunted Mind* and *Two Kinds*, which we began in Chapter 3, using more advanced Python techniques to mine quantitative data from texts.

Part IV: Statistics for the Humanities

Part IV consists of two chapters that introduce key statistical concepts. The significance of understanding statistical ideas and methods cannot be underestimated since they appear explicitly and implicitly in just about every aspect of modern life. Furthermore, statistics is perhaps the most important and effective mathematical toolset for analyzing large datasets, and in particular datasets in the humanities.

Chapter 6, "Introduction to Descriptive Statistics", introduces descriptive statistics, showing what information and insights can be uncovered from relationships between values measured on the same variable. We will start with measures of central tendency – mode, mean, and median – and dispersion measures like variance, standard deviation, and skewness, all explained with practical examples and a critical lens. Then, we will delve into distributions, and in particular the normal distribution, and focus on how it can be used to summarize and generalize data patterns. In the final section, we will combine these statistical measures with Python programming to conduct the first step of a distant reading of four books by two notable American authors, Edith Wharton and Gertrude Stein, whose similar backgrounds provide a unique foundation for comparison.

Chapter 7, "Exploring Relationships Between Variables", introduces the concept of correlation as a way to evaluate the relationships between different variables. Here you will learn to assess connections between phenomena using tools like Pearson correlation coefficient, Spearman's rank correlation coefficient, and the Chi-square test, concluding with linear regression for approximating the correlation between two numerical variables. Throughout the chapter, we will use these measures to analyze the writing of 17th-century playwright Pierre Corneille and the urbanization and industrialization of Germany at the turn of the 20th century, and we further investigate the literary style of Wharton and Stein.

Part V: Introduction to network theory

Part V, the final part of the book, is dedicated to network theory, a mathematical framework that has become central to understanding connections and relationships in various fields like history, literature, and social sciences. However, while networks are widely referenced, misconceptions often arise from misunderstandings of the underlying mathematical principles. Whether networks represent verbal, social, or geographical connections, the core mathematics of network theory remain consistent.

Chapter 8, "Network Theory and Applications in Humanities", introduces the basic concepts of graph theory and their application to various case studies in art, history and literature. We begin with definitions of basic components like nodes and edges and then explore different types of graphs, as well as key properties, including degrees, paths, and connectedness. We illustrate these concepts using examples from diverse fields: the relationships in Shakespeare's plays, family ties between dynasties in 1618 Europe and the educational backgrounds of British architects in the 18th century.

Chapter 9, "Center and Periphery in Network Theory", focuses on the relationships between center and periphery in a network. It introduces various centrality measures, highlighting the strengths and limitations of each, and explores techniques to identify clusters of interconnected nodes and to assess their importance within the network. Additionally, the chapter examines metrics like density, clustering coefficient, and degree distribution to evaluate the overall structure of a network. The chapter's examples cover a diverse selection of networks, including, among others, the Hankyu railway network in Japan, the Inca road network, the Royal Society of the 17th century, the 19th-century Underground Railway, and even the character network of the *Game of Thrones*.

On the programming side, three in-depth case studies will offer you hands-on experience with advanced tools for building and analyzing networks. In the first case study, you'll analyze the European E-road system, a sprawling network connecting cities across Europe since 1950. In the second case, you will uncover the voting patterns between countries in the Eurovision Song Contest. The third case study brings you back to our literary study of the works of Edith Wharton and Gertrude Stein, using network analysis and natural language processing tools to examine patterns of noun usage in their texts. These exercises combine data manipulation, network analysis, and computational text analysis, to reveal new insights into both physical and literary networks.

In order to review and practice concepts and techniques introduced in the book, each chapter includes several questions related to the material. The answers to questions marked with an asterisk (*) appear at the end of the chapter.

References

Armandz, Cécile, and Christian Henriot. 2023. "Beyond Digital Humanities Thinking Computationally: A Position Paper". *Beyond Digital Humanities: How Computational Methods are Reshaping Scholarly Research, Sep 2023*, Aix en Provence, France.

Jacob, Sharin R., and Mark Warschauer. 2018. "Computational Thinking and Literacy." *Journal of Computer Science Integration* 1(1). https://doi.org/10.26716/jcsi.2018.01.1.1

Selisker, Scott, 2016. "Digital Humanities Knowledge: Reflections on the Introductory Graduate Syllabus". In *Debates in the Digital Humanities*, edited by Matthew K. Gold and Lauren F. Klein. University of Minnesota Press.

Steen, Lynn Arthur (Ed.). 2001. *Mathematics and Democracy: The Case for Quantitative Literacy*. The National Council of Education and the Disciplines.

Vee, Annette. 2013. "Understanding Computer Programming as a Literacy". *Literacy in Composition Studies* 1(2): 42–64.

1 Sets, numbers, and functions

A set is a Many that allows itself to be thought of as a One.

Georg Cantor, 1884 (Cantor 1932, 204)

No one shall expel us from the paradise that Cantor has created for us.

David Hilbert, 1926 (Hilbert 1926, 170)

The following chapter introduces a number of mathematical concepts, originating from set theory that form theoretical building blocks, which underlie both logical thinking and computer programming. In general, almost all mathematical concepts can be formally defined using the theory of sets. However, it is not our intention to provide an exhaustive survey of definitions and theorems of set theory. Instead, we aim to introduce the most fundamental concepts: sets, cardinality, tuples, and functions.

The chapter is based on the gradual development of concepts, so it is advisable to move on only after you have thoroughly understood the previous concept.

1.1 Set theory

1.1.1 What is a set?

An art scholar might examine a collection of images, a historian could investigate a collection of sources, and a literary scholar analyzes a collection of books or poems. What these various corpora have in common is that they are all **sets** and are, therefore, characterized by properties defined in set theory, the field of mathematics, which will be described subsequently.

A set is a logical object, considered today as the cornerstone of mathematics. In particular, sets form the basis for the structure of databases and functions, two entities which will be discussed throughout this book.

The history of set theory is different than the history of most mathematical fields. The latter are usually the result of a long process of development until a stroke of genius turns them into a formal theory; this moment is often reached independently

DOI: 10.4324/9781003502814-2

or simultaneously by several mathematicians. In contrast, set theory was the result of the work of a single mathematician, Georg Cantor (1845–1918).[1]

Cantor's early work on number theory did not reflect his pioneering and revolutionary future work. Following an 1872 visit to Switzerland in which he encountered Richard Dedekind (1831–1916), the German mathematician, he began working on ways to formally express the infinity of numbers. The outcome was a short article, entitled "On the Properties of all Real Algebraic Numbers"[2] (*Ueber eine Eigenschaft des Inbegriffes aller reellen algebraischen Zahlen*), published in 1874 in August Carl's *Journal für die reine und angewandte Mathematics* (Journal of Pure and Applied Mathematics). Although it was only four and a half pages long, it marked the beginning of set theory. In his article, Cantor defined sets in general, and described ways of comparing different sets. His most radical result was proving that there are differing types of infinities, some of which were larger than others (some of these infinities will be described in this chapter).

Leopold Kronecker (1823–1891), the editor-in-chief of the journal, opposed the "wild" and abstract use of sets in Cantor's article, and even postponed Cantor's professorship candidacy at the University of Berlin in fear of his corrupting effect on the youth. Due to this controversy, his 1878 follow-up article led to a power struggle between two members of the journal's editorial board; Kronecker opposed Cantor's new publication, while Karl Weierstrass (1815–1897), often cited as the "father of modern analysis", came to his defense. Eventually, Cantor's supplementary article was published in spite of the editor's reservation.

Despite the complexity of the issues with which Cantor dealt and the controversy his work created; the concept of a set is quite simple. Informally, it can be described as a "collection of things" without any significance to order or repetition. Anything can be a member of a set: a word, a number, an idea, a living creature, and even another set.[3] Often, members have something in common but there is no necessity for such rationale; a set can just be a random collection. Objects belonging to a set are called **elements**. By writing $x \in A$, we indicate that a particular element x belongs to a set A. By writing $x \notin A$, we indicate that a particular element x does not belongs to a set A.

1.1.2 How to define a set

The simplest way to define a set is to explicitly specify all the elements belonging to it, usually listing them between curly brackets (braces), {}. For example, the following set, {Oscar Wilde, Umberto Eco, Elsa Morante, Mikhail Bulgakov} contains names of authors. The set {a,e,i,o,u} is a set of English vowels, while the set {А, О, У, Ы, Э, Я, Ё, Ю, И, Е} contains Russian vowels. As mentioned previously, order and repetitions are not significant. Therefore, the set of English vowels is both {a,e,i,o,u} and {a,a,a,i,o,e,e,u}. A set does not have to contain many elements: even if it has just one element, such as {15}, it is a set for all intents and purposes.

This method of describing sets, as simple as it may be, is not always the most convenient way. An explicit enumeration of all the citizens of the United States will take too long. A set containing all the people who ever lived will take even longer, and the set of all the positive odd numbers, {1, 3, 5, 7, …} will never end. In such cases, it is best to describe sets according to the common attributes of their elements. For example, the set of all the citizens of Brazil will be formally described as {x| x is a citizen of Brazil}. This can be translated to the following sentence: "x, such that x is a citizen of Brazil". Similarly, {x | x is an odd positive number} denotes the set of all odd and positive numbers.

Questions

1. *Which of the following sets contains all the letters of 'shoe'?
    a. {h,s,e,o}
    b. {x | x is a letter in "hoses"}
    c. {x| x is an English letter}
    d. {s,e,o,e,h}
2. *Which of the following is true?
    a. $1789 \in \{x|a\ revolution\ began\ in\ year\ x\}$
    b. *Hamlet* $\in \{x|x\ is\ a\ play\ written\ by\ Shakespeare\}$
    c. $f \in \{A, B, C, D, E, F\}$

1.1.3 Russell's paradox

Sets can usually be defined using an explicit record of its elements or a specific condition, determining which elements belong to them. Originally, set theory defined sets as any definable collection. However, Bertrand Russell (1872–1970) discovered in 1901 that there are sets that cannot be described by these methods. He presented this problem through a paradox, which has been since dubbed as "Russell's paradox" (Ernest Zermelo discovered this paradox two years earlier but did not publish his finding).

This paradox has different versions, and one of them is the barber paradox. Assume that in a certain town lives a male barber who shaves all the men who do not shave themselves and only these men. If so, the question arises: is the barber supposed to shave himself or not? If he does not shave himself, then by the definition of his role, he must shave himself. But, if he shaves himself, then by the definition of his role, he is not supposed to shave himself. Therefore, the existence of such a barber is logically impossible.

In Russell's version the paradoxical set was a set whose members are all the sets that do not belong to themselves. Much like the barber from the previous paradox, it is impossible to determine, without contradiction, whether this particular set belongs to itself or does not belong to itself. This paradox led to the development of an "axiomatic set theory". For our purposes regarding sets related to the humanities, we do not have to deal with Russell's paradox and can adopt a simpler and more intuitive definition of sets. This type of set theory is often called "naive set theory".

1.1.4 Relationships between sets: equality, inclusion, and disjointedness

Different sets, when they relate to each other, can be equal, contained, or disjoint.

We say that two sets are **equal** if they have the same elements. For example, the set {"Mother", 3, "Charles de Gaulle"} and the set {3, "Charles de Gaulle", "Mother"} are equal because they only differ in the order of their elements. Another example is the equality between the following two sets:

A = {"English", "Mandarin"}
B = {x| x is one of the two most spoken languages in the world in 2022}.

Although each set is defined differently (A is explicitly defined, while B is defined by a condition), there is no difference between the actual elements belonging to both sets. As a result, these sets are equal.

Containment (or inclusion) is a hierarchical relation between two sets. We say that set A is *a* **subset** of set B, and therefore is **contained** in it, if all the elements of A also belong to B. We can alternatively say that B is a **superset** of set A. We express this relation with the symbol $\subseteq$, e.g., $A \subseteq B$.

For example, the set {Amos Oz, Eshkol Nevo, Yehudit Katzir} is contained in the set of Hebrew authors, since each of its members – Amos Oz, Eshkol Nevo, and Yehudit Katzir – is also a Hebrew author.

A relation of inclusion between two sets does not rule out equality between those sets, since all the members of a given set belong to that same set. For example, $\{a,b,c,d\} \subseteq \{a,b,c,d\}$ because each one of the members of the subset (a,b,c, and d) also belong to the superset.

Questions

3. *Given the following three sets:
   A is the set of all natural numbers.[4]
   B= {2,4,6}
   C= {2,3,4,6}

   Which of the following is true?

   a. $A \subseteq B$
   b. $A \subseteq C$
   c. $B \subseteq C$
   d. $A \subseteq A$
   e. $C \subseteq A$
   f. $B \subseteq A$

4. *For A = {"Modern Times", "The Great Dictator"}, define a subset of A, which is not equal to it, and a superset of A, which is not equal to it.

Disjoint sets are sets that have no common elements at all. For example: the set of all positive numbers and the set of all the works of Shakespeare are disjoint.

Of course, there are many sets that have other degrees of overlap and cannot, therefore, be defined as either equal, disjoint or contained in one or another of the sets. For example, the relation between the set of all verbs in English and the set of all English words that begin with the letter M is neither of the three afore-mentioned relations. It is quite clear that the sets are not equal, because the word 'write' belongs to the first set but not to the second one. They are not disjoint, since 'match' is both a verb and a word beginning with M. In addition, none of the two sets is contained in the other, because each set has words that do not belong to the other set.

The set of elements belonging simultaneously to both sets A and B is called the **intersection** of these sets and is marked $A \cap B$. For example, $\{a,b,c\} \cap \{a,b,d\} = \{a,b\}$.

Questions

5.   *Let A and B be equal sets. What does $A \cap B$ equal to?
   a. A new and smaller set
   b. A
   c. Unknown
6.   *Let A and B be sets such that $A \subseteq B$. What does $A \cap B$ equal to?
   d. A
   e. Unknown
   f. B
7.   *Let A and B be disjoint sets. How many elements are in $A \cap B$?
   g. 1
   h. 0
   i. Unknown

The **empty set** is a set that does not contain any elements. This set is marked with the Norwegian letter $\varnothing$ as suggested by the French mathematician Andre Weil (1906–1998) in 1939. This set can be expressed as {} but can also be expressed as the set of all Russian presidents born before 1820, or as the set of all female American presidents. The fact that a set has a clear condition defining its members does not require that any such elements exist. Campbell (2016) expressed this concept in her poem "Set Theory":

> There is only one empty set, and it exists around the world, the set of loyal lovers and living unicorns, honest politicians and the contents of the intersection of any two non-intersecting circles.

Following the definition of inclusion, we can claim that the empty set is a subset of any set, since all of its (non-existent) elements also belong to the superset. Imagine a bag full of different objects; this is a set. A subset is a set that is formed when some of the objects are removed from the original bag. This makes it easier to understand why the empty set, in which there are no objects at all, is formed from any set following the removal of all its elements.

Defining sets, or finding subsets, is not just an abstract mathematical matter, it is a practical necessity. For example, constructing a database requires an understanding of the type of data that will belong to this database. In other words, we have to define the set to which all the potential objects belong to, prior to the actual construction of the database. Thus, only a proper pre-design of a historical archive will enable the mutual storage of textual, visual, and auditory documents (although, unfortunately, historical archives are rarely pre-designed). Another example is that of filtering computational operations, such as filtering rows in a computerized spreadsheet. These are actually a subset selection from a larger set (the set of all rows in the case of a spreadsheet).

1.2 Cardinality of sets

The issue that troubled Cantor and eventually led to the development of set theory was how to calculate the size of a set, or at least compare different sets. This magnitude is called the **cardinality** of a set, and is marked with two vertical lines, similar to the marking of an absolute value. The cardinality of a set A is marked $|A|$.

When the number of members is finite, the cardinality is simply the number of members in that set. For example, $|\{2,4,6\}| = 3$.

Questions

8. *Let $A = \{1,2,3,4,5,6\}$ and $B=\{x|x$ is an English letter$\}$. Calculate the cardinality of both sets.
9. *Calculate, |set of all letters in the word 'hello'|

But most sets are not finite, they are infinite. In fact, there is a huge variety of sets of infinite power, such as the set of all integers (customarily marked $\mathbb{Z}$), the set of all real numbers (marked $\mathbb{R}$), or the spectrum of colors in human perception.

Continuous as the stars that shine
And twinkle on the milky way,
They stretched in never-ending line
Along the margin of a bay:
Ten thousand saw I at a glance,
Tossing their heads in sprightly dance.
      William Wordsworth, "I Wandered
          Lonely as a Cloud" (1807)

"I Wandered Lonely as a Cloud" (also known as "Daffodils") describes the experience of looking at a field of daffodils stretching toward the horizon, constituting a "never ending line", at least as far as the viewer can perceive. But what did Wordsworth mean? Let's rephrase, what do we mean when we talk about infinity? Why does it feel intuitive to link a never-ending field of daffodils to the Milky Way, but not to the real numbers that densely fill the real line without any gaps?

To answer these questions, we return to Cantor, who proved that there are several types of infinite sets. We define them all as infinite because counting their elements cannot be completed. However, as he proves, there are greater and lesser infinities. The two basic types that fit our needs (Cantor proves that there are an infinite quantity of different infinite cardinalities) are **countably infinite sets** and the *continuum*.

The field of daffodils and the Milky Way, even if they continued indefinitely, are composed of discrete points. After choosing a starting point, we could, theoretically, begin a systematic counting of the set, even if this process would never end. Such sets are similar to the set of all integers or the set of all natural numbers, both of which can be represented as distinct points along the number line. Such cardinality is called **countably infinite** and marked with $\aleph_0$.

In comparison, other infinite sets are dense in such a way that prevents any process of counting. In sets like the set of all real numbers or the set of points on a plane, on which the common system of axes is drawn in math classes, there is no gap between one point and another. This type of cardinality is called the **continuum** and is marked $\aleph$.[5] A painting, unlike a text which is a finite set of words, is an example of an infinite continuous set of points.

It is important to note that this intuitive explanation is not exhaustive since there are infinite sets that appear to us to be dense, such as the rational numbers, even though they can be depicted as discrete. Imagine the rational numbers as pairs of whole numbers and draw them on two axes. The resultant image would be distinct points in space. The magnitude of this set is $\aleph_0$.

Questions

10. *Determine which of the following sets is finite, countably infinite or of the cardinality of the continuum.
    a. Words written in 19th-century journals.
    b. All possible combinations of Hebrew letters.
    c. The timeline of human experience.
    d. The range of human emotions.
    e. The range of colors in a painting.
    f. Place names in an atlas.
    g. Points on a map.

Human experience is infinite in its essence. When one stands at a certain point and decides which direction to turn, they are faced with an infinite number of options. Each option is an angle between 0 and 360 degrees. But why limit ourselves to whole units? After all, one can also turn half a degree, a quarter of a degree and so on. Nevertheless, sources in the humanities are almost always discrete, and almost always finite. Whether they are spoken words or written texts, a painting or a movie, scholars of the humanities focus on distinct points and the connections between them. As a result, mathematical fields that deal with continuous infinite sets, such as differential and integral calculus, are more difficult to directly associate with the humanities than fields that deal with discrete sets, such as statistics or network theory, which we will focus on in this book.

1.3 Cartesian product

As we stated previously, sets are collections of objects whose order has no significance. Nevertheless, often, order is necessary for defining our data. For example, if we know that the protagonist of Tolstoy's novel, *Anna Karenina,* is called "Anna Karenina", we can infer that her forename is Anna and her surname is Karenina. This reasoning is based on our knowledge of how names in Western culture are ordered. On the other hand, if we know that the protagonist of Cao Xueqin's book, *Dream of the Red Chamber*, is called "Jia Baoyu", it should be clear to anyone who is familiar with Chinese culture that his forename is Baoyu and his surname is Jia, since the order is reversed in this culture.

The next two sets, {Elaine Thompson-Herah, Shelly-Ann Fraser-Pryce, Shericka Jackson} and {Shericka Jackson, Elaine Thompson-Herah, Shelly-Ann Fraser-Pryce}, are identical, because their elements are the same. Both sets include the names of the first three runners in the 100-meters women's final at the 2020 Tokyo Summer Olympics. However, sports fans would like to know the order in which these three runners reached the finish line. The fact that Elaine Thompson-Herah came first, Shelly-Ann Fraser-Pryce second, and Shericka Jackson third is essential in determining the result of the race. A different order would have led us to the conclusion that this is a completely different race.

To give meaning to order in set theory, we introduce the concept of an **ordered pair**. Such a pair is usually denoted by round parentheses, (a,b); a is the first coordinate of the pair and b is the second coordinate. The pair (a,b) is not equal to the pair (b,a) unless a=b, just as "King John", the English king between 1199 and 1216, is not the same as the name "John King", the American news anchor of CNN. We can similarly define a triple, a quadruple, or a quintuple. In general, we will define an n-tuple as an ordered list of n objects, in which the order of the elements is of importance.

Questions

11. *Given that the following are n-tuples, what is n in each case?
    a. (1,2,3)
    b. (Rasputin, is, best, known, for, his, role, as, a, mystical, adviser, in, the, court, of, Czar, Nicholas, II, of, Russia.)
    c. The address on the envelope, depicted in Figure 1.1

Figure 1.1 An example of an addressed envelope.

    d. The address on the American Civil War envelope (to Mrs. Ruthann Smith) depicted in Figure 1.2

Figure 1.2 An envelope from the American Civil War.

e. The notes of *Ode to Joy* depicted in Figure 1.3

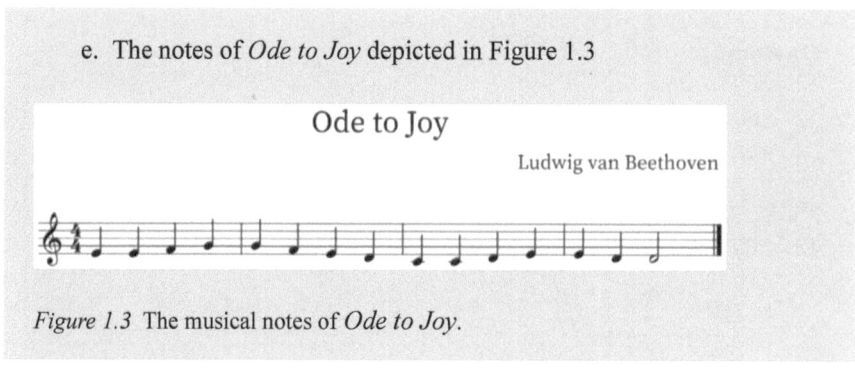

Figure 1.3 The musical notes of *Ode to Joy*.

These examples show that the elements in n-tuples do not have to be of similar types. For example, an address will include a street name, a house number (perhaps the number of an apartment), the name of a locality, a zip code, and a name of a country. The n-tuple, which we call an address, has a mixture of names (of differing meanings) and numbers. We understand the address if we know the pre-designed order of these various types.

A **Cartesian product**, named after the French philosopher Rene Descartes (1596–1650), describes an operation between two or more sets, which creates a new set that includes n-tuples, where n is the number of sets involved in the action. We will demonstrate this on two sets, A and B. The Cartesian product AxB includes all the possible ordered pairs in which the left element is from A and the right element is from B.

For example, for A = {1,2}, B = {a, b, c} we get the following Cartesian product:

AxB = {(1, a), (1, b), (1, c), (2, a), (2, b), (2, c)}.

Similarly, BxA= {(a, 1), (a, 2), (b, 1), (b, 2), (c, 1), (c, 2)}

Following the definition of Cartesian products, the order of the n-tuples is determined by the order of the sets in the product. For example, a mail carrier who distributes mail, or an electronic system that sorts the mail items, assumes that the address is structured according to the Cartesian product: {street names} X {natural numbers} X {locality names} X {zip code} X {country names or not at all}. If we change the order of the product, the result will be completely different, and the mail carrier (or the sorting machine) will have difficulty understanding the addresses written on the envelopes.

Note that the sets do not have to be different. For A = {1,2}, the Cartesian product AxA (sometimes denoted as A^2) can be defined as follows: AxA = {(1,1), (1,2), (2,1), (2, 2)}. Although Rene Descartes never used ordered pairs, his inspirational construction of two-dimensional planes, triggered these ideas. Since 1847, the representation of geometric points in a plane as an ordered pair was coined "Cartesian coordinates".[6] Both values were real numbers, so the Cartesian plane, containing these coordinates, is $\mathbb{R}^2$.

Questions

12. *By which of the following signs, do we identify sets?
    a. {}
    b. ||
    c. A
13. *By which of the following signs, do we identify n-tuples?
    a. ,
    b. ()
    c. a
14. *Which of the following defines the set of all birthdates as a cartesian product?
    a. {x|x is a date}
    b. $\mathbb{N} \times \mathbb{N} \times \mathbb{N}$
    c. $\{1,...,31\} \times \{1,...,12\} \times \mathbb{Z}^7$

Defining a set as a Cartesian product is often the first step towards defining data. For example, if we wanted to create a list of countries and their year of independence, we would start by understanding that our resultant piece of data is an ordered pair where the first element is a country and the second is an integer. Only then can we approach the creation of the list itself, in which each country will be matched to the appropriate number.

1.4 Functions

Functions are a familiar concept, which most people study in their secondary education. However, schools customarily link functions to calculus, rather than teaching what functions are. In fact, real-valued functions (those that receive real numbers and return real numbers) are merely a particular case of functions; there are many other functions that have nothing to do with numbers.

It is almost impossible to understand functions without relating them to sets. Arthur Moritz Schönflies (1853–1928), one of the pioneers of set theory, noted in 1900: "The development of set theory had its source in the effort to produce clear analyses of two fundamental mathematical concepts, namely the concepts of argument and of function" (McLarty 1988, 83).

A function is a matching of elements from one set, known as the **domain**, to elements from a second set, known as the **range**. Such a relation requires that each element in the domain is matched with a single element in the range. For example, every Italian citizen has a single ID number, so a function can be defined by matching each person to the corresponding ID number. The domain in this case will be the set of Italian citizens, and the range will be the set of all natural numbers. On the other hand, it is impossible to define an inverse function between the natural numbers and the set of Italian citizens, because not every natural number can be used as an ID number of an Italian citizen. Such a matching between numbers and citizens would leave many numbers without a counterpart.

Questions

15. *Figure 1.4 illustrates various matchings between two sets using a directed graph (see Chapter 8, for definition). Identify which of the following are functions in this figure.

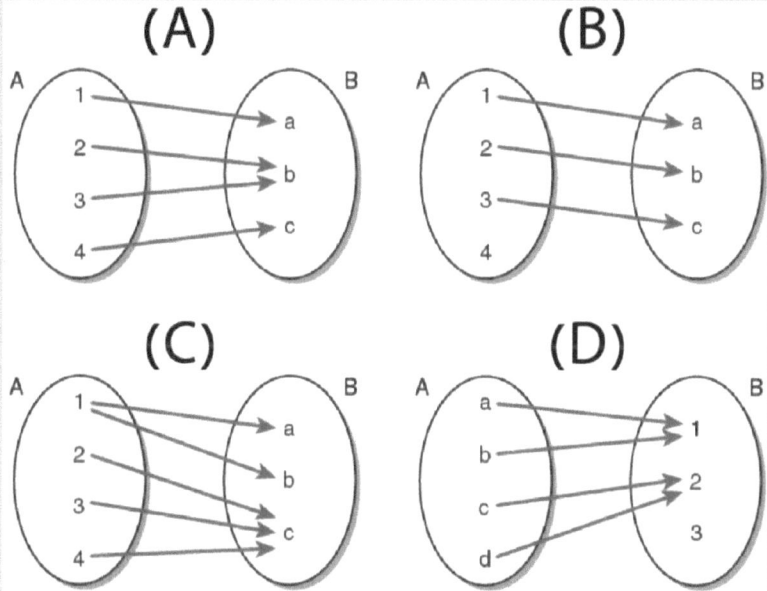

Figure 1.4 Four matchings between sets A and B.

It's important to clarify that there is no reason to assume that all possible relations satisfy the conditions of a function. If we try to match elephants to the food they eat, we will not get a function, because elephants eat more than one thing. Similarly, a parent–child matching is not a function, because not every parent has a single child; the parents of several children will be matched with each of their children, as reflected in a family tree illustration. Furthermore, if we assume that the domain and range are sets of people, we could also have people in our domain who have no children at all, and therefore have no match.

However, we could define an appropriate function between people and their set of children (not their individual children). The domain will still be a set of people (because we match people with something else), but the range will be a set of all sets of people (because the result of this particular function is a set of people). This is a function, because each person would be matched with a single set of children (and those that have no children would be matched with the empty set). Every element in the domain is matched with a single element in the range.

Formally, if we denote the domain by X and the range by Y, the function creates a set of ordered pairs (x, y) so that each x in X is matched with a single element y in Y. For example, if we go into a store, we can assume that each product has a fixed price. Therefore, we can define a function that runs from the set of all products in the store (domain) to the set of all non-negative numbers (the range cannot include negative prices) resulting in ordered pairs of (price, product). These ordered pairs are what we see in the final bill we receive at the cashier of the store.

1.5 Fuzzy sets

The rationale behind different approaches to processing and storing information is often linked to different ways of defining and operating on sets, and yet we sometimes encounter ambiguities in our concepts or definitions. Set theory is not meant to be counter-intuitive but rather to formally define concepts that we understand intuitively in our daily use of language. The formality of set theory makes it possible to avoid ambiguities, and as a result describe complex relations. However, vague phrases like "It is hot today", "This is a successful student", or "This product is expensive" are very common and are impossible to decipher using set theory. Adjectives such as 'hot', 'good', and 'expensive' have no clear boundaries, and their definitions depend on the speaker's perceptions and feelings. For some people a score higher than 80 will be considered good, while for others, a score of 60 will be more than enough. Therefore, it is difficult to define the set of all people who received a high score, or the set of hot days, unless it is categorically determined what is a high score and what is a hot day.

In 1965, an engineering professor named Lotfi A. Zadeh (1921–2017) proposed a mathematical solution for similar ambiguities in the world of engineering. Zadeh defined **fuzzy sets**, in which belonging is not an absolute quality but a proportionate one, which might even be partial. In a 1992 interview, Zadeh claimed that his goal was much greater than simply solving problems in engineering:

> I expected people in the social sciences – economics, psychology, philosophy, linguistics, politics, sociology, religion and numerous other areas to pick up on it. It's been somewhat of a mystery to me why even to this day, so few social scientists have discovered how useful it could be.

(Lair 1994, 46)

Despite his expectations, many people in the social sciences and humanities are unfamiliar with the theory of fuzzy sets.

Classical set theory is based on a clear and unequivocal definition of membership in a set. This type of definition creates a binary distinction between those who belong to the set and those who do not belong. In contrast, the idea behind fuzzy sets is that belonging to a set can be partial, and therefore should be defined according to a degree of belonging. In this vein, a value of 1 reflects complete membership to a set, values close to 1 (e.g., 0.8 or 0.9) will indicate a strong but partial affiliation, values smaller than 0.5 but larger than 0 will indicate that these members are more "external" than "internal", and are therefore weak members of the set, while 0 indicates that the respective object does not belong to the set at all. Fuzzy sets combine a quantitative and qualitative assessment of set belonging.

The theory of fuzzy sets can represent the uncertainty we regularly find in the evaluation of human emotions, in the analysis of a historical narrative, or in the examination of literary and linguistic texts. Therefore, fuzzy sets can provide models to describe the complexities of the world without necessarily defining categories as merely binary.

Questions

16. *What are the pros and cons of defining a set as fuzzy?
17. *Which of the following sets would you define as fuzzy?
    a. A set of English letters
    b. A set of marks in manuscripts we identify as letters
    c. A set of voices we identify as words
    d. A set of musical tunes
    e. A set of musical notes

Glossary

Cardinality The number of elements in a set. It is denoted by |A| for a set A.
Cartesian Product The set of all ordered pairs from two sets A and B, denoted $A \times B$.
Continuum A set with infinite elements that cannot be enumerated, such as the real numbers, where no gaps exist between elements.
Countably Infinite A set with infinite elements that can be enumerated, such as the set of natural numbers.
Disjoint Sets Two sets that have no elements in common.
Domain The set of all possible inputs (or arguments) for a function.
Element An individual object within a set. If an object x is an element of set A, it is denoted as $x \in A$.
Empty Set A set that contains no elements, denoted as $\varnothing$.
Function A relation between sets where each element in the domain is associated with exactly one element in the range.

Fuzzy Set A set where membership is a matter of degree rather than an absolute yes or no. Members can partially belong to the set.

n-tuple An ordered list of n elements. A 2-tuple is called an ordered pair.

Range The set of all possible outputs for a function.

Set A collection of distinct objects, typically considered without any particular order or repetition. Elements of a set can be anything, such as numbers, words, or ideas.

Subset A set A is a subset of B if all elements of A are also elements of B. This is denoted as $A \subseteq B$.

Notes

1 Some argue that the idea of a "set" can be traced to the writings of the Roman philosopher Porphyry of Tyre (234–305) (Ferreirós 2019).

2 Real numbers are the combination of rational and irrational numbers. They can be intuitively understood and constructed as lengths of straight segments or their corresponding negative number. This set is denoted using the symbol $\mathbb{R}$.

3 According to Bonnie Joe Campbell (2016) in her poem "Set Theory":

> You can fill your sets with cardinal numbers or with cardinals,
> and those cardinals might have red feathers or army green
> feathers or might be cardinals in the Vatican voting for a pope.

4 Natural numbers are positive whole numbers. The set of all natural numbers is usually denoted $\mathbb{N}$.

5 In this book, we will not deal with the continuum hypothesis, which asserts that there are no sets whose cardinality is strictly between $\aleph_0$ and the continuum. The truth or falsity of this hypothesis cannot be proven.

6 The first use of this phrase was in Hamilton (1847, 217). The phrase Cartesian product was first used in the mid-1930s.

7 $\mathbb{Z}$ denotes the set of all integers.

References

Campbell, Bonnie Joe. 2016. "Set Theory". *Mississippi Review* 43(3): 132–133.

Cantor, Cantor. 1932. *Gesammelte Abhandlungen*, Abraham Fraenkel and Ernst Zermelo, eds. Springer Verlag.

Ferreirós, José. 2019. "The Early Development of Set Theory". *The Stanford Encyclopedia of Philosophy* (Summer 2019 Edition). https://plato.stanford.edu/archives/sum2019/entries/settheory-early/.

Hamilton, William R. 1847. "On Quaternions; Or on a New System of Imaginaries in Algebra". *The London, Edinburgh, and Dublin Philosophical Magazine and Journal of Science* 31(207): 214–219. https://doi.org/10.1080/14786444408645047

Hilbert, David. 1926. "Über das Unendliche". *Mathematische Annalen* 95: 161–190. https://doi.org/10.1007/BF01206605

Lair, Betty. 1994. "Interview with Lotfi Zadeh: Creator of Fuzzy Logic". *Azerbaijan International* 2(4): 46–50.

McLarty, C. 1988. "Defining Sets as Sets of Points of Spaces". *Journal of Philosophical Logic* 17: 75–90.

Solutions to selected questions

1. The correct answers are a, b, and d.
2. The correct answers are a and b.
3. The true statements are c, d, e, and f.
4. This question has many possible answers. A possible subset of A is {" Modern Times"} and a possible superset of A is {x|x is a movie directed by Charlie Chaplin}.
5. The correct answer is b.
6. The correct answer is a.
7. The correct answer is b.
8. |A|=6, |B|=26.
9. |set of all letters in the word 'hello' |=4
10. a and f are finite. b is countably infinite. c, d, e, f, and g have the cardinality of the continuum.
11. These are the correct n values: a – 3, b – 20, c – 7, d – 4 (Northridgeway, Orleans, C.C., N.Y.), e – 15.
12. The correct answer is a.
13. The correct answer is b.
14. The correct answer is c (assuming a European date format).
15. The following are functions: A and D.
16. There is no single answer to this question. Fuzzy sets are closer to our use of categories as indefinite and ambiguous. However, they do not provide the same certainty as classical sets.
17. The correct answers are b, c, and d.

2   Algorithmics

This chapter deals with solving problems, but not just any problems – only those that can and, in many cases, should be solved using computer software. We will focus on problems related to the humanities. Although our discussion will ultimately focus on concepts from computer science such as algorithms, input, and output, let's go back to the pre-computer era and think about how we approach complex tasks. One useful strategy that is often employed is called "divide and conquer". According to this strategy, complex tasks are divided into smaller more manageable tasks, which are then "conquered" individually, and whose solutions are combined to produce a solution for the original task. We will illustrate this by considering the complex task of compiling a concordance.

2.1 Solving complex problems

2.1.1 Concordances and their creation

A concordance is a compilation containing a list of words that appear in a particular text or group of texts. The most well-known concordance is the concordance of the Bible, in which all the words that appear in the Bible are presented in alphabetical order, and next to each word the references where the word appears are indicated, sometimes also with the verse itself (Figure 2.1).

Creating a concordance in the pre-computer era was a difficult, lengthy, and expensive project, and as a result, only masterpieces such as the Bible, the Quran, or the writings of important writers such as Shakespeare or James Joyce were deemed worthy of concordances. The first known concordance is that of the Vulgate (Latin translation of the Bible), which was compiled in the 13th century by the French Dominican friar, Hugh of Saint-Cher, with the help of 500 monks.

A particularly famous and important concordance is the Hebrew Concordance to the Bible, commonly called *Heichal HaKodesh* (literally: the holy temple), compiled by Dr. Solomon Mandelkern, and first published in 1896. Mandelkern spent over 20 years working on crafting this concordance, with two main purposes: correcting the many errors and omissions he found in previous works,

DOI: 10.4324/9781003502814-3

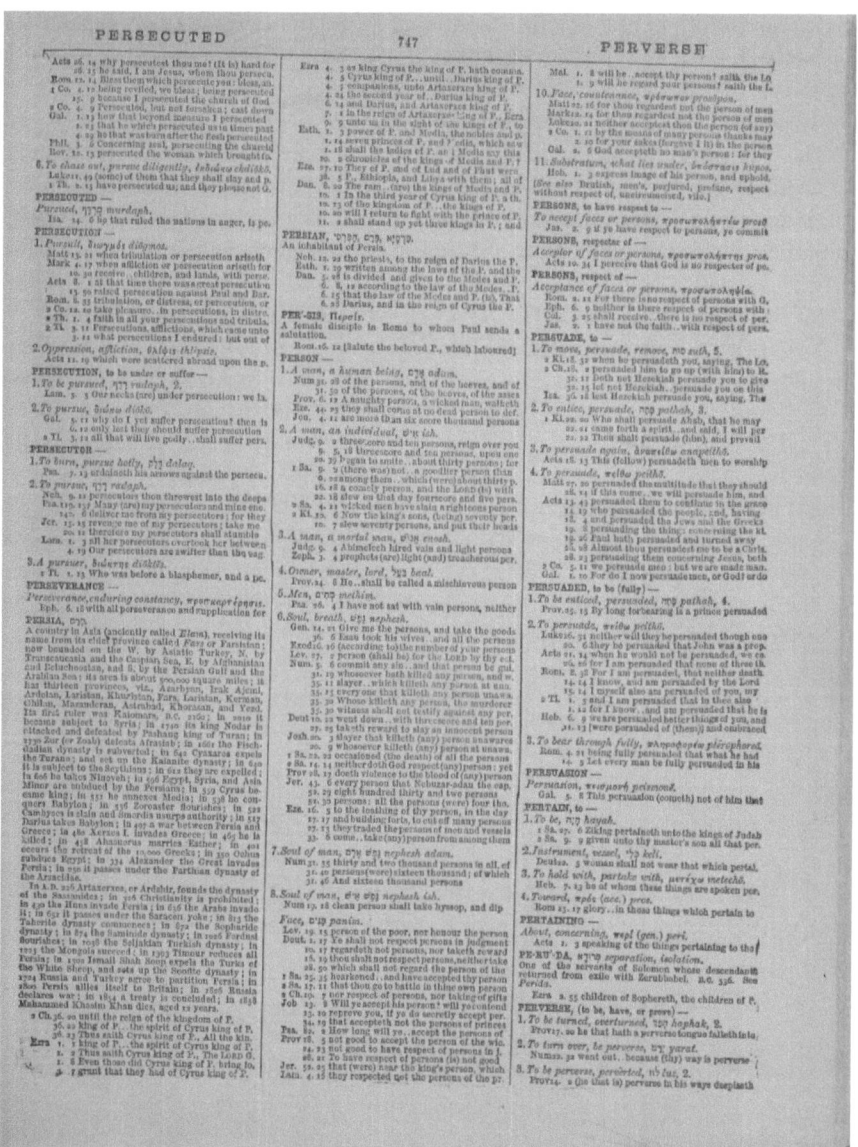

Figure 2.1 An exemplary page from the Analytical Concordance to the Bible.

Source: Young (1882, 643).

and adding entries for all prepositions, pronouns, and proper nouns. Evidence of the enormity of the task can be found in an excerpt from a letter by the poet and writer, Frischmann (n.d.), in which he describes Mandelkern's life work.

> It happened in Leipzig a few years ago. The workshop was at 4 Querstraße , and the protagonist, Solomon Mandelkern, was at work there. For several months, while I was staying in Leipzig, I would visit his home almost every day. It was enough to look at the man's face to know what his work meant to him. His body, which had once been healthy and robust, became gaunt and emaciated, and his cheeks, which had once been full, became sunken and hollow. A strange kind of fire was already evident in his eyes, like the fire we sometimes find in the eyes of people whose minds are clouded and destined to go mad. And this is not surprising: for over twenty years the man was engaged in one task into which he invested all his mind and all his strength; for over twenty years, he worked and labored on it every day, putting in around twenty hours. Sometimes, after he had already gone to bed at night and a word would suddenly come and tap his brain or a misplaced period would come and shake his strings, he would get up and put on his coat again and return to his desk – that man was a hero. That man did a small thing: he created for us a Hebrew and Aramaic concordance, the likes of which have never existed before – it is the concordance he called *Heichal HaKodesh.*

How would you have approached the task of creating the concordance if you had been in Mandelkern's place at the end of the 19th century?

Questions

1. Try creating a mini-concordance for the first two verses in the Bible (from the King James Version). Describe the process precisely.

    [1:1] In the beginning God created the heaven and the earth. [1:2] And the earth was without form, and void; and darkness was upon the face of the deep. And the Spirit of God moved upon the face of the waters.

2.1.2 A pre-computer method for creating a concordance

The method we suggest here uses cards, dividers, and a card catalog box, in which we place the cards in order. On each card, we will list the references of a particular word form. The dividers will group together cards associated with the same uninflected base form (also referred to as **lemma**).[1] We will include in our concordance only **content words** (i.e., nouns, verbs, adjectives, and adverbs) and ignore frequent **function words** (e.g., articles, conjunctions, and prepositions). Moreover, we will assume the text is short, so that one catalog box is enough.

To create the concordance, we will go over the text word by word, and for each one we will perform the following sequence of actions, divided into three main steps:

1. Word filtering
   a. If the word is a function word (e.g., a, the, and, or, to, of, is):
      i. ignore and continue to the next word.
2. Finding/creating the **divider**
   a. If the word is a base form:
      i. look for a divider with the word.
   b. Otherwise (i.e., the word is a complex word):
      i. extract the base form from the complex word.
      ii. look for a divider with the base form.
   c. If a divider with the base form does not exist – we will create a new divider for it and place it in its location in the box.
3. Finding/creating the **index card** and updating it
   a. Check if there is a card with the original word after the base-form divider.
   b. If there is no card:
      i. create a new card.
   c. Copy the verse and its reference to the card.
   d. Place the card in its location in the box, after the divider.

Questions

2. Execute the method described above for the following two verses:

   [1:1] In the beginning God created the heaven and the earth. [1:2] And the earth was without form, and void; and darkness was upon the face of the deep. And the Spirit of God moved upon the face of the waters.

The following list shows the result of executing this method. The uppercased words indicate dividers, and the italicized ones indicate cards. Instead of quoting the full verse, we only list its number.

BEGINNING, *beginning* [1:1]
CREATE, *created* [1:1]
DARKNESS, *darkness* [1:2]
DEEP, *deep* [1:2]
EARTH, *earth* [1:1], [1:2]

FACE, *face* [1:2], [1:2]
FORM, *form* [1:2]
GOD, *God* [1:1], [1:2]
HEAVEN, *heaven* [1:1]
MOVE, *moved* [1:2]
SPIRIT, *Spirit* [1:2]
VOID, *void* [1:2]

Simply attempting to create a concordance for two verses, comprising 39 words of which only 16 have actual content, offers a glimpse into the challenges of Mandelkern's task. To get a sense of scale, it is worth noting that the main part of the concordance *Heichal HaKodesh* contains about 1,250 pages, not including separate sections for the function word 'asher', which appears about 5,000 times in the Bible, as well as Aramaic words and proper names. In total, the concordance spans 1,532 pages. There is no doubt that the manual process of creating such a concordance was labor-intensive and required linguistic expertise, meticulousness, systematicity, and a great deal of time.

2.1.3 Creating concordances in the computer age

The computer age brought with it many changes in relation to concordances. Evidence of the transformation can be found already in 1959 in a report written by Ze'ev Ben-Haim, a linguist and founder of the Historical Dictionary of the Hebrew Language Project:

News, that concordances are being created by machines at an unprecedented speed in the United States, reached me, by hearsay, before my journey to England ... And now, Busa [Roberto Busa – an Italian priest, a pioneer in using computers for linguistic and literary analysis] shows that a concordance of the works of Thomas Aquinas, whose vocabulary is estimated at 13 million words, can be prepared in four years or less.

(Ben Haim 1959, 115)

In the 1950s people got to see how the process of creating concordances turned from a life's work into a project that lasted only a few years. Today, due to the accelerated technological development, there is no longer any need to compile and print a bulky concordance like *Heichal HaKodesh*. You can easily and quickly retrieve a concordance for a specific entry, or more precisely, to find all its occurrences in any text that exists in digital format. One such example is the online concordance of the King James Bible: https://thekingsbible.com/Concordance/.

In this book we will not learn to write concordance software, but rather how to design solutions for similar problems. We will also experiment with writing short codes that solve sub-problems related to computerized processing of language data.

2.2 Computational solution to problems

2.2.1 Algorithms

The process we described for creating a concordance is actually an algorithm. The term **algorithm** denotes a general recipe for solving a problem – a well-defined sequence of actions, which, when executed, can reliably attain the desired outcome. The word was born out of the mispronunciation of the name Muhammad al-Khwarizmi, a Persian mathematician, who lived over 1,000 years ago, and is considered to have first formulated the systematic rules for addition, subtraction, multiplication, and division of decimal numbers.

Algorithms describe a process that starts with the problem data – that is, the **input,** and ends with the desired result – that is, the **output.** Part of the definition of an algorithm includes a description of the set of possible legal inputs, as well as a definition of the desired output for each input.

At the design stage, algorithms are written in the form of **pseudocode**, that is, in plain language, not in a programming language. However, as can be seen in the previous example, the steps required to solve the problem are formulated concisely, orderly, and systematically, addressing the different cases that may occur (e.g., lack of divider). After the algorithm is written this way, the translation of the pseudo-code into programming code is relatively straightforward.

Writing algorithms requires a process of **abstraction**, that is, distilling the fundamental properties of the data, while ignoring other properties that are not relevant. The goal is to present a general solution that can be applied to different types of input. Thus, for example, the concordance creation algorithm does not refer in any way to the content of the words it handles. It also does not refer to the number of words that appear in the text. In fact, the same algorithm applies to the first two verses of the Bible as it does to the entire Bible. Also, there is no reference in the algorithm to the text on which the concordance is based – any text is suitable.

An algorithm includes another type of abstraction. It presents a general outline for solving a problem but does not go into details that are essential to accomplishing the task. For example, with regards to the concordance creation algorithm presented above:

* Do we distinguish between function words and content words?
* Do we group complex words (i.e., suffixed and prefixed forms) under a shared base form, or do we list each form separately?
* How do we determine the base form of a complex word?
* How do we search for a divider or a card?

The first three questions relate to the linguistic aspects of the process. The concordance creator needs to decide which words to include and which to ignore. Highly frequent functions words such as 'a' and 'of' are often omitted. Furthermore, the distinction between "complex words" and "base forms" is not always a trivial matter. In English, inflected words, such as plural nouns or verbs in the past tense, are typically created by appending a suffix (-s for plurals, -ed for past tense) at the end of the base form. Consequently, these words appear together when sorted alphabetically, even when they are not grouped under one base form. In Hebrew, on the other hand, base forms often appear with prefixes which translate to independent English words like 'the', 'and', and 'to'. Consequently, (complex) prefixed forms can only appear together if they are grouped under a shared divider associated with the base form. In both cases, the identification of the base form requires natural language processing software for conducting morphological analysis and disambiguation.

The fourth question, the divider search question, is of a different kind. Unlike the first three questions, whose answers affect the final product, with regards to searching for a divider, as long as the search process works properly (i.e., if a divider exists, it is found), the search method is insignificant. As we will see in the next section, there are different methods for searching for an item in a list, but the choice of any particular method is not the "concern" of the concordance creation algorithm. This "divide-and-conquer" approach calls for **modularization**. For the concordance creation algorithm, each of its sub-processes (finding the base form, searching for a divider, etc.) is handled by a separate **module**, or "black box" which receives input (e.g., a word) and returns the required output (e.g., the base form associated with the word). The mode of operation is not the "responsibility" of the general algorithm.

2.2.2 Searching for items in lists

Search is a very common operation in computerized systems. Nowadays, the term "search" is almost automatically associated with internet search engines, however here we will deal with a different kind of search – searching for an item in a list.

As mentioned, the process of searching for a divider is a separate module within the concordance creation process. We will focus on it and examine how it is executed by computer programs. To do so, we will treat the dividers in the card catalog box as a list of words arranged in ascending alphabetical order. For simplicity, we will consider the first ten words: 'beginning', 'create', 'darkness', 'deep', 'earth', 'face', 'form', 'God', 'heaven', 'move'.

The algorithm will receive a word as input and will return as output an indication of whether the word is found in the list, or in our terms – whether a divider for that word exists in the card catalog box.

In computer science, a list with items arranged in ascending or descending order is called a **sorted list**. Each item in a sorted list has one position where it can be found – the item before it is smaller than it (in alphabetical order) and the item after it is larger. If the item is not in this position, then it is not in the list. We will ignore for now the situation where items appear more than once in the list.

A computerized search for an item in a list is a computational problem, and as with any computational problem, we first define the input and output.

In this case, the input consists of a sorted list (L) and the item we are looking for (x).

The output is an indication of whether the item is found in the list or not. In computer science, such an indication, which can be either one of two values, True or False, is called a **Boolean** value.

The word *Boolean* is derived from the name of George Boole, a 19th-century English mathematician and philosopher, one of the founders of modern logic. Boole created Boolean algebra, which is the basis for computer science.

Simple linear search

We will start with the simplest algorithm, which checks if some item exists in a list. The algorithm is called **linear search** because it goes over the list items according to their location, one after the other.

We will go through the list item by item:
    For each item:
            If the item equals x:
                return True and terminate.
    If we reached the end of the list (x was not found):
        return False.

It is often convenient to describe algorithms using flowcharts. Let's look at the flowchart of the algorithm we suggested. (See Figure 2.2.)

The process described in the flowchart is serial – the actions are carried out one after the other according to the direction of the arrows. At certain points in the process, denoted by diamond shapes, **conditional branching** occurs – one arrow indicates the direction of progress if some condition holds, and another arrow indicates the direction of progress if the condition does not hold. The entry and exit points of the algorithm appear as ellipses.

The algorithm is built as a **loop** with one starting point and two exit points, where the output value (True or False) is returned. The loop goes through the items one by one, comparing each item with the input value x. Each sequence of transitions

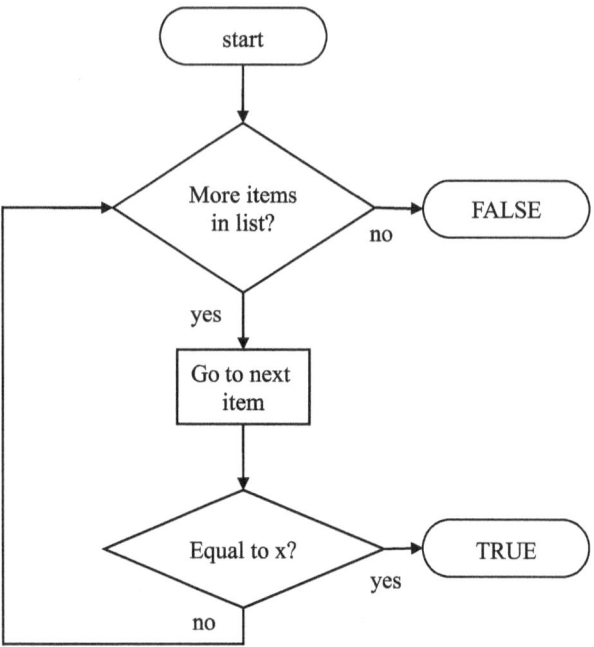

Figure 2.2 Simple linear search flowchart.

within the loop is called an **iteration**. The process reaches the exit points, and terminates in two cases:

- The end of the list is reached, i.e there are no more items in the list (the upper diamond decision node);
- An item equal to x was found (the lower diamond decision node).

Questions

3. What will happen if x appears twice in the list?

One important challenge in formulating such an algorithm is handling all possible cases, especially boundaries and exceptions. For example, in our algorithm, the command "go to next item" (in the middle rectangle) is executed only after we verify that there still are items in the list (in the upper diamond above it). Naturally, this is an unnecessary step for humans, who would not examine the value of an item in an empty list. But when computer programs try to perform operations on undefined positions and values, unpredictable and even disastrous consequences may follow. Therefore, when we formulate an algorithm, we must think about the process from the computer's point of view. These boundary cases are known vulnerabilities in the craft of programming.

Improved linear search

When examining algorithms in computer science, one important measure that distinguishes between algorithms is **efficiency**. An algorithm's efficiency is measured, among other things, by estimating the number of operations that will be required to perform the task in relation to the amount of data, especially in the worst-case scenario. In our case, we will estimate the algorithm's efficiency by counting the number of comparisons between list items and x.

Questions

4. With respect to the linear search algorithm and the sample list ('beginning', 'create', 'darkness', 'deep', 'earth', 'face', 'form', 'God', 'heaven', 'move'):
    a. What choice of "x" (word) would be the best-case scenario in terms of number of comparisons?
    b. What choice of "x" (word) would be the worst-case scenario in terms of number of comparisons?

It is very convenient to use a concrete example, but when we evaluate the efficiency of an algorithm, we will not refer to a specific case, but rather to the general case. For instance, we will claim that the best-case scenario is when x is equal to the first item in the list; after one comparison we will return True and terminate. The worst-case scenario is when x is equal to the last item in the list or does not appear in it at all. In this case the number of required comparisons is equal to the number of items in the list. Such a claim is valid, regardless of the length of the list.

A significant disadvantage of the very simple linear search algorithm lies in the fact that it is inefficient.

Before you continue reading – think how the algorithm's efficiency could be improved.

Since our list is sorted, there is only one position where item x can be found. Once we pass that position, there is no point in continuing to traverse the rest of the list. That is, the **termination condition** we set in the first algorithm is not optimal.

See Figure 2.3 for the improved algorithm and its flowchart.

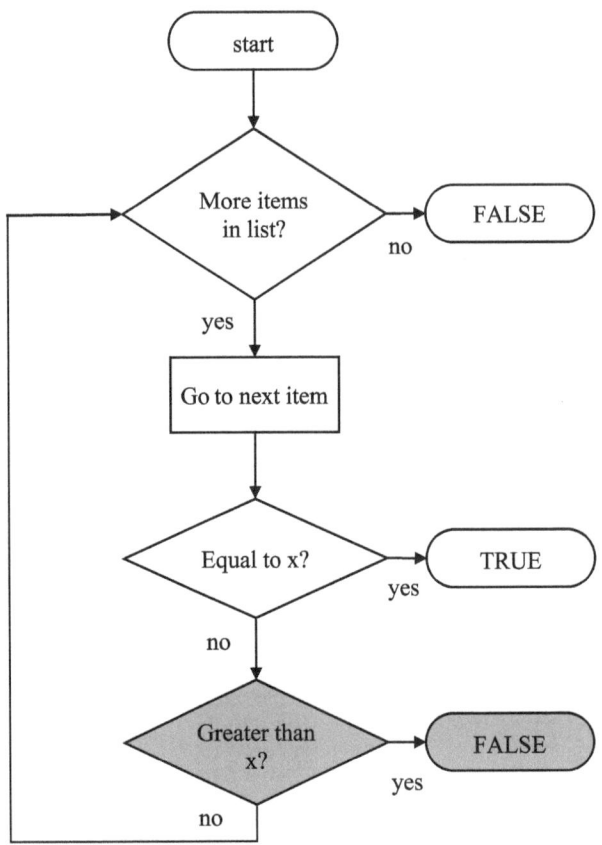

Figure 2.3 Improved linear search flowchart.

We will go through the list item by item:
  For each item:
   If the item equals x:
    return True and terminate.
   Otherwise:
    If the item is greater than x:
     return False and terminate.
  If we reached the end of the list (x was not found):
   return False.

In the improved algorithm, another loop termination condition was added (the shaded shapes). Now there is no need to perform comparisons between x and items that by definition cannot be equal to it.

The improved linear search is more efficient for any input that does not appear in the list and is not greater than the last item in the list. In such cases, the process will stop before reaching the end of the list. Thus, for example, if we use the improved linear search when searching for "banana" in the sorted list of fruits "apple, mango, strawberry" we will stop after the second comparison, and will not need to continue comparing. However, if we search for "watermelon" we will need to go through the entire list before we conclude that it is not found, regardless of the algorithm.

The two algorithms are equally efficient when the searched item does appear in the list. They both stop when the item is found. Another case is when the input equals the last item in the list. In both algorithms, the same number of comparisons will be made, equal to the list length.

Although the difference in efficiency between the two algorithms is insignificant in the case of a 10-item list, when a software requires multiple searches and extremely long lists are involved, reducing the number of operations can affect performance. Computer science experts strive to develop optimal algorithms that will work efficiently even under difficult conditions. In the next section, we will explore an even more efficient algorithm for searching lists: binary search.

Binary Search

The algorithm is called **binary search** because it is performed by splitting the list (or, more accurately, the search range) into two parts at each step. As with the improved linear search, this algorithm also assumes that the list is sorted.

See Figure 2.4 for a description of the algorithm using pseudocode.

---

1. Set the search range as the entire list.
2. Perform the following steps as long as the search range is not empty:
   Find the middle item of the range (for an even range – the item before the middle):
   If x equals the middle item:
   >   return True and terminate.
   Else:
   >   If x is smaller than the middle item:
   >   >   reduce the search range to the left (smaller) half and return to step 2.
   >   Else:
   >   >   reduce the search range to the right (larger) half and return to step 2.

3. If we got here, it means the search range is empty. Return False.

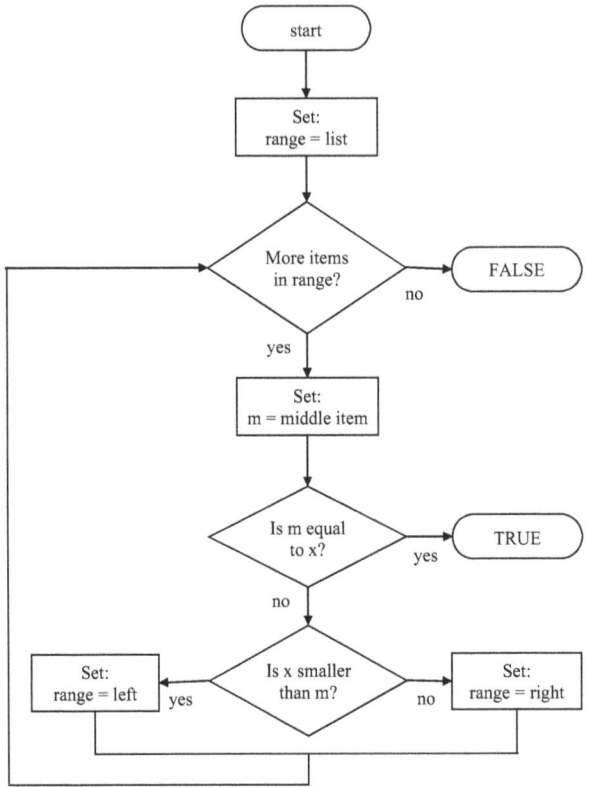

Figure 2.4 Binary search flowchart.

As with linear search, binary search has a central loop that is executed as long as the termination conditions are not met. Similarly to simple linear search, there are two termination conditions here: x is found, or the search range is empty. However, binary search is a much more efficient algorithm than linear search, and this becomes clear the longer the list is. Recall that algorithm efficiency is measured relative to the worst-case scenario; in this case, when no items are left in the search range.

We will illustrate the efficiency of the algorithm using an example. Suppose that the number of items in the list is 32, and we assume the worst-case scenario: the value of x is smaller than the smallest member in the list (that is, the first member, since the list is sorted). In this case, the search will continue until there are no items left in the search range.

In the following diagram, each dash represents an item in the search range in an iteration, from top to bottom, starting from 32 items in the first iteration, up to only one item in the sixth and last iteration.

```
32   – – – – – – – – – – – – – – – – – – – – – – – – – – — — — — — – – –
16   – – – – – – – – – – – – – – – – –
8    – – – – – – – –
4    – – – –
2    – –
1    –
```

At each step, or iteration, we compare x with the middle item in the list (schematically indicated by shading). The maximum number of comparisons is actually the number of times you can split the range in half. In the case of an initial list of 32 items, after 5 unsuccessful comparisons we will reach a range with one item, and in the worst-case scenario this item will not be equal to x. That is, we will make 6 comparisons in total.

As mentioned, the efficiency of an algorithm is evaluated in relation to the amount of data (N). In this specific example N=32, and the maximum number of comparisons is 6. In general, for any N, the maximum number of comparisons is $\log_2(N)+1$. The difference in efficiency between the linear and binary algorithms grows very rapidly (exponentially). When the list length is 100, the maximum number of comparisons in binary search is 8 (versus 100 in linear search).

Questions

5. *Let's compare the improved linear search with the binary search using our example: 'beginning', 'create', 'darkness', 'deep', 'earth', 'face', 'form', 'God', 'heaven', 'move'.
    a. Give an example of an input for which linear search is more efficient. What is the difference between the two algorithms in terms of the number of searches?
    b. Give an example of an input where binary search is more efficient. What is the difference between the two algorithms in terms of the number of searches?

2.2.3 Graphs and networks

The Seven Bridges of Königsberg

Our discussion of graphs and networks begins with a story.

Königsberg is a city in current-day Russia, now called Kaliningrad. In the early 18th century, it was a prosperous commercial city. As can be seen on the map

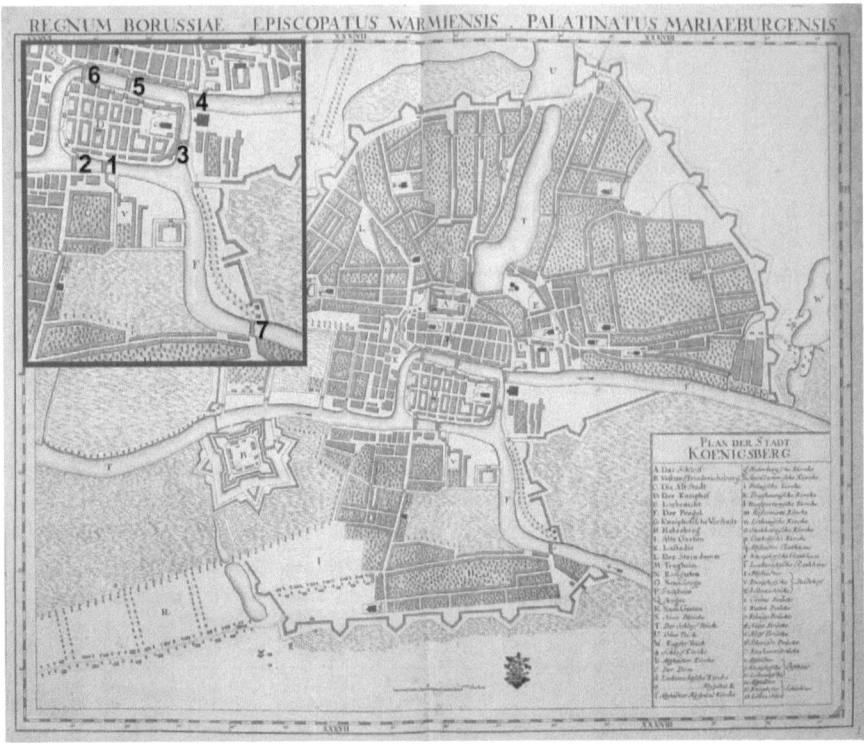

Figure 2.5 The Seven Bridges of Königsberg.

Source: Suchodolec (1763).

(Figure 2.5), a large river, the Pregolya, flows through the heart of the city, dividing it into several areas, including the island in the center of the map. The land areas are connected by seven bridges. Today, the Seven Bridges of Königsberg are particularly famous for being the name of a well-known mathematical puzzle.

The story begins in 1735. Leonhard Euler, who was a Swiss mathematician, astronomer, and engineer, was asked by the residents of Königsberg to solve a problem that had preoccupied the residents for generations: is it possible to find a path that crosses all seven of the city's bridges without crossing any single bridge more than once. At first Euler argued that he, as a mathematician, had no advantage in finding a solution. The answer, he claimed, depends solely on logic. However, he continued to think about it, and in the process invented an analytical method, which developed into the field of mathematics now known as **graph theory**.

Euler's problem-solving process was based on **abstraction**.

First, Euler noted that the layout of the streets crossing the city has no impact on the solution; the only thing that matters is how the land areas are connected to

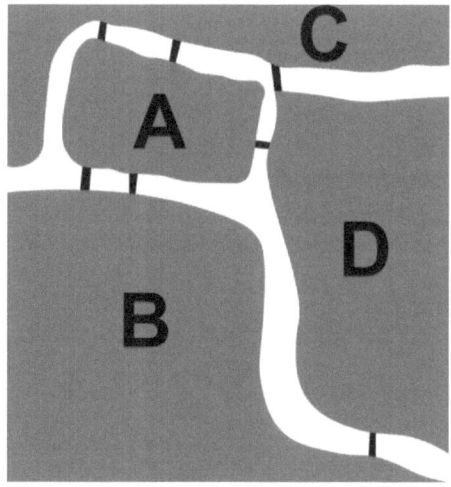

Figure 2.6 The Seven Bridges of Königsberg – abstraction.

each other. As such, he chose to treat each land area as a single unit. He created a schematic version of the map, in which each area is represented by a single Latin letter (Figure 2.6).

The more abstract representation helped him focus on the essence. He decided to represent possible paths using the letters. Thus, for example, the path BAC denoted crossing two bridges – the bridge between B and A and the bridge between A and C. A representation of a path crossing all seven bridges would include eight letters. In principle, Euler could have tried to go through all the possibilities in Königsberg itself, and check if there is a path that meets these conditions; but he sought to find a more general solution.

The next step Euler took was the one that introduced a new and revolutionary mathematical entity into the world. He proposed to represent the system of seven bridges as a network described by a **graph** consisting of **nodes** and **edges**. Each node in the graph represents one of the land areas, and each edge represents one of the bridges (Figure 2.7). A solution to the bridge problem, if one exists, would be expressed as a path passing through all nodes via all edges, traversing each edge only once.

Using the graph Euler proved that there is no solution to the problem of the seven bridges. But far beyond providing a specific solution to the problem that troubled the residents of the city, Euler's general solution paved the way for a new way of thinking about problem solving.

Since Euler's time, a lot of water has flowed under the bridges of Königsberg. In 1875 the residents decided to build an additional bridge between B and C. Whether by design or by chance, with the addition of the bridge it became possible to find a path crossing all bridges once. However, unfortunately, not all bridges survived.

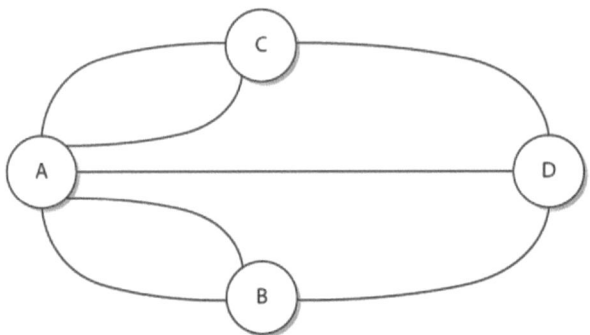

Figure 2.7 The Seven Bridges of Königsberg as a graph.

Towards the end of World War II, in 1944, British planes bombed the city, resulting in two of the bridges being destroyed. Later in the war, the old city was almost completely destroyed. Over the years, two more bridges were demolished to make way for a highway. After its annexation to the Soviet Union, the city was renamed Kaliningrad, but the Seven Bridges of Königsberg were immortalized forever in the history of mathematics.

The Königsberg bridges problem and Euler's solution are a prime example of a problem-solving process, and in particular the abstraction required to start from a concrete example, with all its features and details, and arrive at a general solution, applicable not just to the specific example we started with. In Euler's case, solving the problem created a central theory in mathematics, but usually we will be satisfied with finding a good and efficient solution.

Searching graphs

Graphs, such as the one Euler devised to solve the Königsberg bridge problem, are an excellent tool for representing networks of interconnected entities: each member of the network is represented as a node, and relationships between members are edges between nodes. One particular type of a network that can be represented with such graphs is social networks. In social networks, such as Facebook, people are connected to other people through symmetric friendship connections (if A is friends with B, then B is friends with A). One of Facebook's capabilities is to suggest new friends to its users, based on mutual friends.

In graph theory terms, the existence of a mutual friend of two members is manifested in the fact that there is a **path** of two edges and one node that connects the nodes representing the two members. In this case, it is said that the **distance** between the two members is 2, or, in other words, the **shortest path** between them spans two edges.

One algorithm that is used to find the shortest path between two nodes in a network is called **Breadth-first Search** (BFS). The BFS algorithm is a general algorithm that is used for systematically traversing nodes in a graph or network. The search is called "breadth" because it progresses in "layers" – it begins with a particular node and scans all its immediate neighbors. From there it continues to nodes that are two edges away from the starting point, 3 edges away, and so on.

Another algorithm for searching a network is called **Depth-first Search** (DFS). In this algorithm, the search begins at some node and continues moving away from the starting point as far as possible. When it reaches a dead end, it backtracks until it reaches a node from which it can continue.

Let's demonstrate the Breadth-first Search process on a "toy example" of a small social network (see Figure 2.8). Note that the edge lengths have no significance. The numbers in the nodes indicate the order in which the nodes are visited.

The search process can start with any node. In our example we start with Sneezy. The second layer includes Sneezy's three immediate neighbors – Bashful, Sleepy, and Snow White – whose distance from Sneezy is 1. The third layer includes all nodes that are at distance 1 from the nodes in the second layer: Happy, Grumpy, and Doc. The fourth and final layer includes all the immediate neighbors of nodes

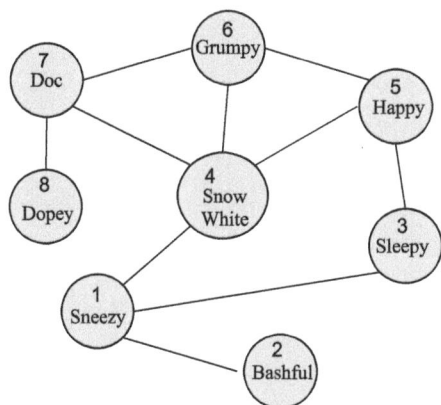

Figure 2.8 Breadth-first Search example.

in the third layer which have not yet been visited. In this case there is only one such node – Dopey.

Finding the distance between two nodes

The breadth-first search process is useful for finding the distance, or in other words, the length of the shortest path, between two nodes. The algorithm traverses the network in a breadth-first fashion, until it finds the target node. At each step, it calculates the distance of a given node from the source node.

However, before we start presenting the algorithm, we should note that the search process makes use of a **queue** data structure. A queue, like a queue in the everyday context, is a list to which new items are added to the end and are removed from the beginning. Any item entering the queue leaves only after all items that were in the queue when it entered have left. In computer science this is called **First In First Out** or FIFO for short.

In breadth-first search, a queue is used to control the traversal order. When nodes are visited and inserted into the queue, they are marked as such to avoid infinite loops. At each step in the algorithm, the first node in the queue is extracted and all its neighbors are scanned. As long as the target node is not found, each neighbor which has not yet been marked as "visited" joins the end of the queue.

Let's define the Breadth-first search algorithm in terms we're already familiar with:

The input is a graph describing a network, a source node (x) and a target node (y).

The output is the distance between x and y, if such a path exists.

Here is the algorithm specified using pseudocode:

Insert the source node (x) at the end of the queue.
   While there are nodes in the queue:
      Extract the first node in the queue (w).
      For each of the immediate neighbors of node w:
         If the neighbor is equal to y:
            distance = distance (w) + 1
            return distance and terminate
         If the neighbor has not been "visited":
            mark neighbor as "visited"
            Set the neighbor's distance to be distance (w) + 1
            insert the neighbor at the end of the queue.

And you can see the flowchart in Figure 2.9.

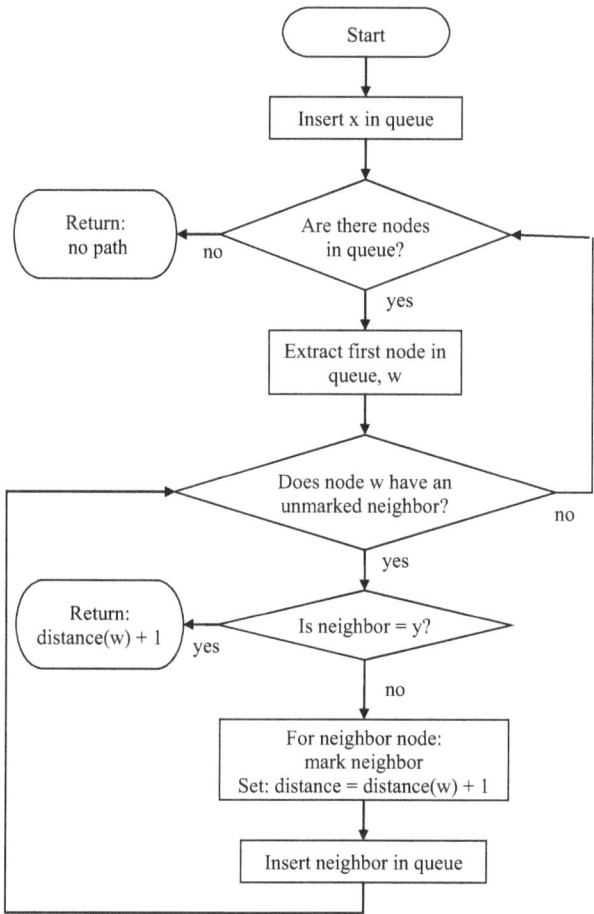

Figure 2.9 Breadth-first Search flowchart.

Questions

6. *According to the algorithm's results – what is the shortest path between Sneezy and Happy?
7. *Are there additional paths of length 2 between Sneezy and Happy?
8. *When there are multiple equal-length paths between source and target, what determines which path is chosen?
9. *Run the BFS algorithm on the network, starting from Dopey, and list the nodes in "layers" according to their relative distance from the starting point.

2.3 Summary

This chapter began with an example – the complex task of compiling a concordance – as an illustration of the strategies that we adopt when we are faced with such a challenge. As we showed, problem-solving strategies are relevant regardless of whether computers are involved in the process or not. We have emphasized that problem solving requires a preliminary stage of **abstraction**, where we extract the features of the problem that are relevant for providing a solution. Part of the solution process also includes "dividing and conquering", or, in other words, defining **modules** that are responsible for performing operations that are "not of concern" to the main process.

In the computational domain, a description of the steps that are required to solve a problem is called an **algorithm**. An algorithm defines a sequence of operations that start with an **input** and produce an **output**. We have seen that algorithms can be described using **pseudocode** or a **flowchart**.

We have identified typical structures in algorithms, such as a **loop** that is executed multiple times, where each "run" is called an **iteration**. As part of defining the loop, we set its **termination conditions**. Another typical structure is a **conditional branch**, which defines two different courses of action based on the input state. Such branches are manifested in conditional statements (if X then Y, else Z).

In the context of the general concordance creating algorithm, we focused on the search module. We learned three different algorithms – **linear search**, **improved linear search**, and **binary search** – which differ in their **efficiency**. However, we saw that efficiency comes at a price – the two more efficient algorithms require the list to be **sorted** (and sorting a list also has its price).

Our discussion on algorithms and problem solving expanded to **graph theory**. We saw how, in an attempt to solve a seemingly geographic problem in the 18th century, Euler invented an important analytical tool– the **graph** – which we still use today, even more so since the advent of internet social networks. The brief overview of graph theory revealed how, through a process of abstraction, it is possible to represent very different domains using the same tool. In the case of the Seven Bridges of Königsberg, the graph, consisting of **nodes** and **edges**, represented land areas, and the bridges connecting them. Conversely, in social networks, the nodes in the graph represent members, while the edges represent friendship relationships between them.

The study of graph theory is not limited to representing problems, but also includes analyzing graphs and finding solutions. A central concept in characterizing connections between nodes (and what they represent) is the **distance** between them, manifested by the length of the **shortest path** from one to the other. In this section, we presented an algorithmic solution for investigating a network. The algorithm, called **Breadth-first Search**, systematically traverses a graph representing a network, and finds the shortest path from the source node to each of the other nodes in the graph. The algorithm uses a **queue** – a list-like data structure described by the term **first in first out** or FIFO for short.

We formulated an algorithm for finding the distance between two nodes by using pseudocode and a flowchart. Formulating algorithms in this way is a very important stage in finding a computational solution to a complex problem. The next stage, which we will deal with in the next chapter, is translating algorithms to a programming language and executing them.

Glossary

Abstraction The process of distilling the fundamental properties of data while ignoring irrelevant properties to present a general solution applicable to various types of input.

Algorithm A well-defined sequence of actions to solve a problem, beginning with input data and ending with the desired output.

Binary search An efficient algorithm for finding an item from a sorted list of items, operating by dividing the list in half repeatedly and comparing the target value to the middle element of the list.

Breadth-first Search (BFS) An algorithm that explores all the immediate neighbors of a node in a graph before moving on to the next level of neighbors. Used for finding the shortest path in graphs.

Concordance A compilation listing words that appear in a text or group of texts, often with references or excerpts indicating the context in which each word appears.

Conditional branch A point in an algorithm where one of two different actions is chosen based on a condition.

Divide and Conquer A strategy for solving complex problems by breaking them down into smaller, more manageable tasks, solving each task individually, and then combining their solutions to address the original problem.

Edge In a graph, an edge is the link that connects two nodes.

Flowchart A visual representation of the steps in a process or algorithm, showing the flow through boxes and connected by arrows.

Graph A mathematical structure used to model relationships between objects. A graph consists of **nodes** (vertices) and **edges** (connections between nodes).

Iteration A single cycle of a repeated process in an algorithm, where the same set of instructions is repeated until a certain condition is met.

Linear search A straightforward search algorithm that checks each element in a list one by one until the target element is found or the list is exhausted.

Loop A programming structure that repeats a sequence of instructions until a specific condition is met.

Node A basic unit of a graph. Nodes are connected by edges.

Pseudocode A description of an algorithm that uses plain language and resembles programming code. It is used to outline the steps required to solve a problem before translating them into a specific programming language.

Queue A data structure in which elements are added at the back and removed from the front, following the First In, First Out (FIFO) principle.

Shortest path The path between two nodes in a graph such that the number of edges is minimized.

Termination condition A specific condition that ends the execution of a loop or a program.

Note

1 In English, the base forms of nouns is their singular form (CAT for 'cat' and 'cats') and the base form of verbs is their infinitive form (TAKE for 'take', 'takes', 'took', 'taking', 'taken').

References

Ben-Haim, Ze'ev. 1959. "The Creation of the Historical Dictionary of the Hebrew Language by the Academy of the Hebrew Language". *Leshonenu* 23(2–3): 102–123.

Frischmann, David. n.d. "Letters on Literature: Letter 9". https://benyehuda.org/read/19045.

Suchodolec, Jan Władysław. 1763. *Regnum Borussiae, Episcopatus Warmiensis, Palatinatus Mariaeburgensis et Culmensis cum territorio Dantiscano et ichnografia urbis Regiomontis*. Unknown Publisher.

Young, Robert. 1882. *Analytical Concordance to the Bible on an Entirley New Plan Containing Every Word in Alphabetical Order*. Dodd, Mead and company.

Solutions to selected questions

3. Since the algorithm stops as soon as an item identical to x is found, an identical item appearing immediately afterwards will not be accessed.

5. Improved linear search vs. binary search:
   a. When the input is the word 'beginning', the improved linear search will only need one comparison, while binary search will first compare 'beginning' with 'earth', then with 'create', and only in the third comparison will find 'beginning'.
   b. When the input is the word 'earth', improved linear search will need to make five comparisons (all the words preceding 'earth') before finding it, while binary search will need to make only one comparison, since 'earth' is the fifth word in the list, just before the middle.

6. Shortest path: Sneezy – Sleepy – Happy

7. Additional paths: Sneezy – Snow White – Happy.

8. At each step the algorithm scans all the current node's neighbors and appends them to the queue one after the other. The order in which they are appended to the queue determines which path among equal-length paths will be chosen. In the case of the path between Sneezy and Happy, the path via Sleepy was chosen because at the stage when all of Sneezy's neighbors were inserted into the queue, Sleepy was inserted before Snow White.

9. Starting from Dopey
   Distance 1: Doc
   Distance 2: Snow White, Grumpy
   Distance 3: Sneezy, Happy
   Distance 4: Bashful, Sleepy

3 Introduction to the Python programming language

A computer program is a sequence of instructions that the computer needs to execute, and the process of writing computer software is called programming. In order for the computer to actually know how to execute the instructions, they need to be written in a pre-defined language that the computer is able to "understand". Such a language is called a **programming language**. Programming languages, like human languages, have syntax – that is, rules that determine how the instructions are written, and semantics – the meaning of the instructions. The vocabulary of programming languages is extremely limited and includes just a few English words called reserved words (for example: 'if', 'while', 'and').

Unlike natural languages, programming languages are unambiguous: each instruction has only one meaning. Another difference between the two types of languages is that computers, unlike humans, cannot cope with errors. So, for example, people have no problem reading and understanding a sentence that contains a misspelled word or a sentence that does not end with a period. If we try to run a program with such errors, the attempt will fail, and we will get an error message. Therefore, when writing computer software, you must be very careful about following the rules.

There are dozens of programming languages, including C, C++, Java, and Perl. In this course we will learn to program in a language called Python, one of the most common programming languages today, both in academia and in the hi-tech industry. Python is considered good for beginners, because it is relatively simple and readable. It is also particularly suitable for working with texts, which makes it useful in the field of digital humanities.

There are different platforms on which you can run Python code. Online platforms like Google Colab and Kaggle Kernels do not require any installation. They are cloud-based solutions, which are accessible from any device with internet connectivity, support collaborative coding, and come with powerful computing resources. Conversely, local development environments such as IDLE, PyCharm, and VS Code need to be installed on your computer and offer robust features for writing and debugging code.

DOI: 10.4324/9781003502814-4

Learning the language and gaining programming experience initially require some effort and patience, but gradually you will develop skills and confidence. As you will come to realize – programming can be very rewarding.

In this unit we will make a first acquaintance with programming in general, and with the Python language in particular. We will learn to write and run simple programs, and also try to handle errors (both intentional and unintentional). At the beginning, the programming process may seem pointless, but even in this introductory chapter we will conduct meaningful textual analyses of literary works. Like swimming, programming is a skill that cannot be mastered just by reading about it. It requires hands-on practice, so you are encouraged to run the code provided in the text and tackle the programming assignments. As we progress in the book, we will see more and more ways in which we can harness the ability to program in Python for collecting and processing data.

3.1 Variables and values

3.1.1 My first program

The tradition in computer science is to start every programming course with a very simple program, called `Hello World`.

The program appears in its entirety in the code box in front of you, and its output in the grey box following it.

```
# This is the traditional first program
print("Hello World!")
```

```
Hello World!
```

The first line starts with the # character, which indicates that what follows is a comment that should be ignored by the software. Programmers use comments to clarify various pieces of code for themselves and others.

The second line features a print function call, used for displaying content on the computer screen. The content appears as an argument inside the parentheses. In our case, the content contains a sequence of letters and other symbols, so it appears between double quotes (but can also appear between single quotes).

When we run this program, the output is displayed on the screen.

Already at this stage we should note that programming requires strict adherence to standard usage of the programming language. Computers, unlike humans, cannot cope with inaccuracies. If you try omitting one of the quotes or one of the parentheses in the code above and then running the program, the result will be an error message.

SyntaxError: unterminated string literal (detected at line 2)

As the message indicates, omitting the parentheses or quotation marks is an error of type SyntaxError; that is, a use that violates the rules of the programming language.

3.1.2 Variables

A key term in the field of programming is **variable**. Variables can be viewed as a kind of a labelled container where information can be stored and modified. In the following code box, the command in the first line demonstrates how we assign a value to a variable called message by using the = sign.

```
message = "Hello World!"
print(message)
```

The arguments which we pass to the function print() can be values ("Hello world!") or variables, as long as they contain some value (e.g., message in the code above). In general, attempting to perform an operation on a variable that has not been assigned a value will result in an error or unexpected behavior. For example, running the code below will result in the message "NameError: name 'text' is not defined".

```
print(text)
```

In fact, the name of the variable is not important; We could use any sequence of letters (with a number of limitations that we will address later). However, it is recommended to use significant variable names, which reflect the nature of the values they contain. This makes it easier to read and understand a computer program when it is more complex.

What do you think will be printed on the screen after you run the following code?

```
name = "Jane Austin"
author = name
  print(author)
```

The ability to assign a variable the values of another variable will be useful to us later.

So far, we have only worked with **string** variables, which store sequences of letters, symbols, and spaces. Another type of variables is **numeric**.

What do you think will be printed on the screen after you run the following code?

```
x = 8
print(x+2)
print(x-4)
print(x*x)
```

After assigning a numerical value to the variable, it is treated as a number for all intents and purposes, and therefore it is possible to execute all standard mathematical operations on it. Also note, that the input to the `print()` function is not limited to string values; Numeric values can also be printed using this function.

The assignment of values to a variable can be done directly (`x = 8`), or by assigning it the value of another variable (`author = name`). Another possibility is to assign a value that depends on the value of the variable itself. We will demonstrate this.

Suppose we count the number of books in our possession, and we currently have 40 books. We can define a variable `number_of_books` and assign it the value 40.

```
number_of_books = 40
```

Note that variable names cannot contain spaces. In their place, it is customary to use an underscore.

If, for example, we purchased a new book and we want to update the value of the variable accordingly, we can do so using the following code.

```
number_of_books = number_of_books + 1
```

If we run this code several times, the value of the variable will increase by 1 each time.

3.1.3 Types and names

As we have seen, it is simple to define variables in Python. You specify the name of the variable and assign it a value using the = operator. So far, we have encountered two types of values: numbers and strings. Each variable can potentially hold any type of value, but as soon as it is assigned one, it "adopts" that value's type. We will illustrate this with the following example.

```
x = "Jane Austen"
x = "8"
x = 8
```

In the three lines of code, different values are assigned to the same variable: strings in the first two lines and an integer in the third. Note the difference between the second and third lines. Although the value in the second line is a number, since it appears between double quotes it is considered a string (with one character). The assignment command on the third line changes both the value of the variable and its type. As we will see later, the type of value stored in the variable determines the actions that can be performed with it and on it.

There are certain restrictions that Python imposes on names that can be given to variables:

- A variable name can only contain Latin letters (uppercase or lowercase) or numbers, as well as symbols such as an underscore (_).
- Python distinguishes between lowercase and uppercase letters (NAME, Name and name are considered three different variables; it is highly recommended not to use such variables in the same program).
- A variable name cannot begin with a number.
- A variable name cannot contain spaces, so it is customary to use an underscore as a separator between words in a variable name that contains more than one word.
- A variable name cannot belong to the reserved words list:

```
and     assert   break class continue
def     del      elif  else  except
exec    finally  for   from  global
if      import   in    is    lambda
None    not      or    pass  print
raise   return   try   while
```

We will get to know some of these words later on.

Questions

1. *Write a piece of code in which two variables are assigned numerical values which represent two page numbers in a book, the first page in any chapter and the last page in the chapter. Print the number of pages in the chapter.

3.2 Logical operators and conditional structure

3.2.1 *Logical operators*

A significant part of programming is based on comparing different values. This comparison is done using **operators.**

Let's focus on the following four operators:

*   < "less than"
*   > "greater than"
*   == "equal to"
*   ! = "not equal to"

Note: It is important to distinguish between =, which is used to assign a value to a variable, and ==, which is used to compare two values. Usually replacing one with the other causes a `TypeError`, but sometimes the program runs and performs different operations than intended.

What do you think will be printed on the screen if we ran the following code?

```
x = 3
y = 9
print(x < y)
print(x > y)
print(x == 3)
print(y != 9)
```

The `print()` function prints either "True" or "False" depending on the operator and the values that are compared. This type of a value is called **Boolean** and it can be only one of two: `True` or `False`.

Note that this code can run successfully only if the variables x and y are assigned a value.

Questions

2. *Write four `print()` statements (different from those in the box above) that print boolean values – two true and two false.

3.2.2 If-else conditional statement

Comparing different values enables us to execute various operations based on the outcome of the comparison. Let's go back to a previous example, where we counted the number of books we own, and after purchasing a new book we updated the counter. Let's say we have room on the shelf for 45 books, and we want to get a message when we need to add a new shelf.

Consider the following code.

```
if number_of_books > 45:
    print("Add a new shelf!")
```

The code assumes that the variable `number_of_books` holds a value. It compares the value of the variable with the number 45. If the variable's value is greater than 45, the `print()` function is executed and it prints the desired message.

What happens if the condition is not met?

The code in the box refers only to the case where the condition is met. When it is not met (i.e. the number of books is less than or equal to 45), nothing happens. To explicitly refer to this situation as well, we will use the reserved word `else`.

```
if number_of_books > 40:
    print("Add a new shelf!")
else:
    print("All OK!")
```

On the one hand, the code seems very intuitive: if the number of books is greater than 40, print one message, otherwise – print another message. However, this is a programming language, and as such, it is important to be precise and to adhere to valid code. Otherwise – the program will not work.

There are two critical constraints that must be respected when writing an if-else statement:

- At the end of the `if` statement and after the `else` there should be colons.
- All commands after the if statement or after the else must appear on the next line, indented to the right.

The examples so far included conditions that compare variables with numeric values, but the operator and conditional structure are also relevant for strings.

```
name1 = "Bonnie"
name2 = "Clyde"
if name1 < name2:
    print(name1)
else:
    print(name2)
```

The relationships "greater than" and "less than" between two strings are determined by alphabetical order. For example, since the letter C appears after the letter B, it is considered greater.

3.3 Functions

Functions are code segments designed to perform some action (or several actions). Python provides various functions called **system functions** one of which is the `print()` function, which is used to print content on the computer screen. When we call this function, we write the content we want printed inside the parentheses that appear after the function name. The values inside the parentheses are called **arguments.** For `print()`, the argument can be any value (number or string), or a variable that contains a value. This function is only the first of a wide variety of functions that we will learn. But before we continue, let's go over some basic terms regarding functions.

3.3.1 Defining functions

In the previous page we discussed a small program which compares the values of two variables and prints the larger one. Let's assume for the sake of demonstration that in the context of a large and complex software we are required to perform this operation several times. Instead of copying and pasting the lines of code in different places in the program, we will define a function and call it every time we need to perform this operation.

The following code defines such a function.

```
def bigger (x, y):
    if x > y:
        return x
    else:
        return y
```

The first line is the definition line of the function. It includes the reserved word def (definition), the name of the function, the **parameters** (here x and y), which are variables that will hold the function's input. When the function is called with two arguments, the values of those arguments are passed to the parameters x and y. A colon obligatorily appears at the end of the line. The name of the function and the names of the parameters are chosen by the programmer, and, similarly to variable names, preferably reflect their purpose.

The following lines contain the body of the function, i.e., the program itself. In our case, the program contains the conditional statement if-else, which compares two variables. However, unlike the previous code, the function bigger() does not print the larger value, but returns it using the reserved word return. In other words – the function accepts two parameters, and returns one value, the larger one.

The function bigger() returns one value, but in general, returning a value is not mandatory for all functions. There are functions that do not return any value at all (like print), and there are functions that return several values.

3.3.2 Calling functions

The function itself, as defined above, does not do anything if it is not called. When calling a function there are two things to keep in mind:

- The input to the function – how many arguments it accepts and of what type.
- The output of the function – how many values it returns and of what type.

Regardless of the function's input and output, the call to the function is always enclosed in parentheses that appear (without a space) after its name. Inside the parentheses are the values that the function receives as arguments, if any.

In the following code segment, the function bigger() is called with two arguments, 1998 and 2002, which are passed directly to the function, without an "intermediary" variable. The function then assigns them as values to its internal variables (x and y), in the order they are passed. The value that is returned by the function (the bigger of the two) is assigned to the variable recent, which is then passed on to the print() function.

```
recent = bigger(1998, 2002)
print(recent)
```

The function we wrote can operate on values of different types. Because of this, the function can also be called with two string values.

```
print(bigger("Barbie", "Ken"))
```

In this code segment, the value that is returned by `bigger()` is passed on directly to `print()`. Such a situation, where the argument of one function is obtained from a call to another function, is called **nesting**.

It is important to clarify that there is no difference between calling a function with values and calling a function with variables. Both ways are valid and are presented here only to demonstrate the different possibilities. However, remember that the use of variables is conditioned by the fact that some value is placed in them. Calling a function with an unknown variable will result in an error.

The example we used here is a bit forced. In fact, there is no great advantage to using the function when it is a simple `if-else` statement. The advantage of defining functions is manifested when the action the function performs is more complex. In such a case, you can write the code that the function executes in one place, and use it when needed, without duplicating it in several places. This way, if you need to modify how the function is programmed, you only have to do it in one place.

This feature is especially noticeable in system functions. When we call the `print()` function, we don't know and don't need to know anything about the code behind it. In addition, even if a new version of Python comes out, in which there will be some change in the way the `print()` function is written, this change will be "transparent" to us.

The ability to divide software into "autonomous" parts is called **modularity**.

Questions

3. *Write a function that receives two numbers and returns their ratio. Make sure to handle illegal cases where the second number is equal to 0. Test the function by calling it with different values.

3.4 Strings

A significant part of research in the digital humanities deals with texts and their analysis. Therefore, in this section we will focus on operations that can be performed on textual variables of type `string`.

3.4.1 Concatenating strings

One basic operation that can be performed on string values is **concatenation**. The operation is performed using the addition operator +.

In the following code, the variable `message` is assigned the value obtained by concatenating the value of the variable `name` to the end of the string "My name is".

```
message = "My name is " + name
```

Note that the concatenation operation is not "sensitive" to the content of the values it concatenates. Therefore, if we want a space character to appear between the strings that we concatenate, we need to add it at the end of the first string (or at the beginning of the second string).

As we have seen, the addition operator + has two different roles:

- Mathematical addition of numeric values (e.g., x + 2)
- Concatenation of strings ("aaa" + "bbb")

The + operator can be used only between values of the same type. If, for example, we try to run the following code, where a numeric variable (`age`) is concatenated to a string ("His age is"), we get a type error message.

```
age = 64
message = "His age is " + age
```

TypeError: can only concatenate str (not "int") to str.

In this case, to fix the error we will use the `str()` function, which takes a number as an argument and returns a string value. This function essentially converts a number type variable to a string type variable, which can then be concatenated to another string value.

```
age = 64
message = "His age is " + str(age)
```

3.4.2 Strings as sequence of characters

Python treats strings as sequences of characters and offers various tools to analyze them. One such tool is the reserved word `in` which we can use to check if a string contains another string.

In the following code box we use `in` to define a function `HasSpace()` that receives a string as an argument and checks whether the string contains a space character. The function then returns a Boolean value, `True` or `False`, accordingly.

```python
def HasSpace(text):
    if " " in text:
        return True
    else:
        return False
```

Note that when we call functions such as `HasSpace()` which return Boolean values we can explicitly compare the function's output to a Boolean value:

```python
if HasSpace(text) == True:
    print("Has space!")
```

Alternatively, we can abbreviate this by omitting the explicit comparison:

```python
if HasSpace(text):
    print("Has space!")
```

And we can also use the reserved word `not` to implicitly compare the value with False.

```python
if not HasSpace(text):
    print("Does not have space!")
```

The function `len()` is used to determine the number of characters in a string. It takes as its argument a string value and returns a numeric integer value, representing the number of characters in that string.

```
author = "Jane Austin"
n = len(author)
print(n)
```

11

In the previous code box we used variables for the string and for the numeric value that is returned by `len()`. The numeric variable was then passed as an argument to `print()`. The same result can be achieved without using any "intermediate" variables. In the next code box, the argument of `print()` is the output of the `len()` function, whose argument is the string itself.

```
print(len("Jane Austin"))
```

The two methods (and other permutations) are equivalent. However, it is often easier to understand code with aptly named variables.

Another tool Python provides for analyzing strings is the use of **indices**, which allow access to individual characters within a string. For example, to print the first letter of the string variable `name` we will run the following code.

```
print(name[0])
```

As you can see, the numbering of characters in a string starts at 0.

If we want to retrieve a sequence of characters within a string, we do so by defining a range of indices.

```
print(name[1:4])
```

The range definition is a bit confusing. The range includes the first index, but not the last index. That is, the range `[1:4]` includes three characters: `[1]`, `[2]`, and `[3]`. For example, if the value of the variable name is the string "Peter", `name[1:4]` is the string "ete".

If we omit the first or last index from the range, it will retrieve the characters from the beginning or up to the end of the string, respectively. So, continuing with the previous example, `name[2:]` is "ter" and `name[:2]` is "Pe".

Questions

4. *Write a short program that defines a string type variable, and then checks the length of the string. If the string length is less than 10, it prints out a message which contains the value of the string and its length, otherwise it prints a message warning that the string is too long.
5. *Write a code segment which calls the function `HasSpace()` with a particular string value, and, depending on the output of the function, prints an appropriate message which includes the string which was examined.

3.5 Loops

A common operation in programming, and in particular, when processing and analyzing texts, is iterating over a sequence of elements, and performing operations on each element. The technical term for this action is a **loop**, whereas each step in the sequence is called an **iteration**.

The command used for this purpose is `for`.

```
for x in "banana":
    print(x)
```

The elements of the sequence in the case of a string are the characters that form it. The `for` loop iterates through the string, character by character, and each time assigns the character as the value of the variable (in our case – x). The number of iterations performed is equal to the number of letters in the string.

Similarly to the `if-else` statement, the command or commands that are executed at each iteration appear indented below the `for` command. Here, too, it is important to adhere to the format:

*   The reserved words `for` and `in`
*   A colon at the end of the for command
*   Indentation ("inner indentation") of the commands within the loop

Note: The `for` loop is one of the only cases in Python where a variable (x in this case) can be used without first assigning it a value.

Let us consider a code segment illustrating the use of `for` loops.

```
num_chars = 0
for x in "banana":
    print(x)
    num_chars = num_chars + 1
print(num_chars)
```

The first line of code contains a command called an initialization command, that is, assigning an initial value to a variable. In this case we initialize the counter `num_chars` before entering the loop. Without this action we would not have been able to use it inside the loop. The `for` loop iterates over the characters in the string "banana", and for each character, it prints it on the screen, and increases the value of `num_chars` by 1. The final line in the code is not indented and is therefore not part of the `for` loop and is executed only after the final character in the string has been accessed.

What do you think the output will be if the command in the last line is also indented? Check it!

Questions

6. *Write a short program that counts and finally prints the number of occurrences of the letter a in the string "abracadabra".
7. *Write a short program that counts and prints the number of words that appear in the string "Life is what happens when you're busy making other plans".

3.6 Lists and list operations

3.6.1 Defining and accessing lists

In the previous section, we explored string variables and values and learned how to concatenate them (+), determine their length (`len(name)`), and extract individual characters and sequences from specific positions in a string (`name[0]` and `name[2:4]`, respectively). In the same context, we also introduced the `for` loop tool (`for...in`), with which it is possible to sequentially access all characters within a string.

In this section, we will explore another type of variable, called **list**.

A list, like a string, contains a sequence of items in a certain order. But while a string contains a sequence of characters (letter, symbols, spaces), a list contains a sequence of items of various types.

In the following code segment, we define two variables with a list type value. The value of lit_words is a list of five strings. The variable mixed_list contains a list of seven items of various types, with two identical items. Regardless of the types of items they hold, lists are defined using square brackets, between which the items appear separated by commas.

```
lit_words = ["poem", "novel", "author", "metaphor",
"simile"]
mixed_list = ["red", 12, "John Lennon", "Russia",
58, "y", 12]
```

Similar to strings, items in the list are accessed using indexes, with 0 indicating the first item.

```
print(lit_words)
print(lit_words[0])
print(mixed_list[2:4])
```

Moreover, the reserved word in can be used to check whether a particular item appears in a list.

```
new_word = "poem"
if new_word in lit_words:
    print("The word " + new_word + " already exits in
    the list.")
```

Like strings, for loops can be used to iterate through lists, item by item. In each iteration the loop variable is assigned the value of the current item in the list. The program below iterates through the lit_words list and counts the number of items in the list that are found in the string text.

```
text = "The author's most famous poem is highly
metaphorical."
lit_count = 0
for word in lit_words:
    if word in text:
        lit_count = lit_count + 1
print(lit_count)
```

3

Although each item in `lit_words` is a word, the condition `if word in text` checks whether the sequence of letters in `word` appears in the text. For this reason, the string `author` in `author's` and the string `metaphor` in `metaphorical` are counted, together with `poem`.

3.6.2 Basic operations on lists

Python provides two types of functions: **general system functions** such as `print()`, `len()`, and `str()`, and **methods**. Methods, like functions, are programs that are designed to perform specific operations. Similarly to functions, methods also accept arguments and return values.

There are two main differences between functions and methods. First, methods are defined for a specific type of variable. The call to the method is made only on variables of the appropriate type for that method. In the next sections we will explore methods specific to strings, as well as those particular to lists. The second difference is reflected in how methods are called.

Counting

In the previous section you were asked to write a program that counts the number of occurrences of a certain character within a string. To do so, you wrote a `for` loop that iterated through the string, character by character, and compared the current character to the character you were asked to count. In fact, the purpose of the task was for you to gain experience in writing `for` loops. You may not be happy to hear that the program you wrote can be replaced by the method `count()`.

Look at the following code segment:

```
text = "abracadabra"
num_char = text.count("a")
```

Notice how this method is called, and how it differs from the way system functions are called. The call to the method `count()` is made on a string type variable. Note that there is a period separating the variable name from the method name. The argument of `count()` is the character whose number of appearances you wish to count. The method returns as a value the number of times the character appears in the string, and in our code, the value is assigned to the variable `num_char`. If such a character is not found, the method returns 0.

The same method can also be used with variables of type list. The following code counts the number of items identical to the number 12 in the `mixed_list` variable.

```
num_item = mixed_list.count(12)
```

Similarly, instead of iterating over a string and counting the number of spaces, we can call the count() method with an blank space as an argument.

```
text = "Life is what happens when you're busy making
other plans."
print(text.count(" ") + 1)
```

Note that in this code segment, the argument of the print function includes a call to the count() method, and the addition of 1 to the returned result (the number of spaces is always 1 less than the number of words).

Appending and removing items

While the method count() can be applied to both strings and lists, some methods are type-specific. Following are the methods that perform the two most basic operations that can be performed specifically on lists:

append()   Add an item to the end of a list.
remove()   Remove an item from a list.

To add an item to the end of a list we call the append() method on the list variable. The value of the new item is passed as an argument to the method.

```
lit_words.append("story")
```

The append() method only operates on list variables that have already been defined. The following code initializes a list variable by assigning it the value of an empty list.

```
colors = []
```

Similarly, it is also possible to remove an item from the list, using the remove() method, when the value of the item, in this case "novel", is passed as an argument.

```
lit_words.remove("novel")
```

Note that since lists allow for multiple items with the same value, the append command (above) can be executed repeatedly, and each time an element (with the same value) is added to the end of the list. The `remove()` method removes the first item whose value corresponds to the value of the argument. If there is no such item, an error message is printed.

Questions

8. *Write a function `IsLit()` which receives a string value and checks whether the string contains at least two items from the list of literary words `lit_words`. If it does it returns the value True, and if not – False. Test the function with different values.
9. *Write a program that iterates over a list of words and creates a new list which includes only the long words (length > 5) of the original list.

The `IsLit` function in Question 8 demonstrates a process called "text classification". In our example, we checked for the presence of words typical of literature as a method for classifying the text. The function was designed to classify a text as literary if at least two words from a predefined list of literary terms appeared in it. While the example in the exercise is a "toy example" based on only six terms, this methodology is commonly applied on a much larger scale.

3.7 Text preparation

In this section we will focus on one of the most central tasks in the digital humanities research program: text processing. We will first learn different Python tools which can be used to take a text and prepare it for performing data analysis.

3.7.1 From text to list

Up until this point we learned that texts can be stored as values of string type variable. We saw that by using a `for` loop we can iterate through the text, character by character. Moreover, we learned how to use the method `count()` to count the spaces between words, as a way to calculate the number of words that a text contains. This solution did provide an answer to the question, but as we will see now, and also as we progress in this textbook, a more convenient representation of text is one in which the relevant units are words and not characters. That is, we will convert a text representation from a string of characters to a list of words.

At the current stage, to facilitate the presentation, we will use short texts. Later on, we will apply the techniques presented here to analyse longer texts.

Let's take as an example a quote from the book *Winnie the Pooh*.

```
quote = " 'What day is it?', asked Winnie the Pooh.
'It is today,' squeaked Piglet. 'My favorite day,'
said Pooh."
```

The quote contains a short dialogue, with each saying by Pooh and Piglet appearing between single quotation marks. However, the entire text appears between double quotation marks, so that Python treats it as a string, containing, among other things, single quotation marks.

Suppose we want to check how many times the word 'day' appears in the text. Do the programming tools we have learned so far support this?

In the previous section we counted the number of occurrences of the letter 'a' in the string 'abracadabra' using a `for` loop that iterated over the string, character by character, and compared each one to the letter a. When the character was the same, we increased the value of the counter by 1. At the end of the loop, the counter held the number of occurrences of the letter.

This algorithm is not sufficient to count the number of occurrences of a specific word; this task requires a much more complex algorithm. But, as mentioned, if we convert the string into a list of words, we can perform such operations easily.

`Split()` is a method that applies to strings and returns a list of the words in the string. In the following code segment this list is assigned to the variable words. The value of the string variable `quote` does not change following the operation.

```
words = quote.split()
```

The value of the list variable `words` is:

```
[" 'What", 'day', 'is', "it?',", 'asked', 'Winnie',
'the', 'Pooh.', "'It", 'is', "today,'", 'squeaked',
'Piglet.', "'My", 'favorite', "day,'", 'said',
'Pooh.']
```

Note the difference between `quote` and `words`. In both cases the words appear in the same order, but in the case of the list `words`, the words appear between single quotation marks, with a comma separating them.

What do you think will be printed on the screen by the following code segment?

```
print(quote[1])
print(words[1])
```

Now that the text we want to analyze is represented by a list of words, we can start extracting basic quantitative data about it. First, we will calculate the number of words in the text.

The function `len()`, with which we previously found the length of a string, can also be used for finding the number of items in a list, or, in our case, the number of words in a text.

```
num_words = len(words)
```

To know how many times a given item appears in a list, we can use the `count()` method, which operates on string-type and list-type variables. This, however, is not as straightforward as it would seem.

Consider the following code.

```
num_w = words.count("day")
```

If you try to run this code you will find that the value that is returned is 1, when the word in fact appears twice (and the string appears three times). Can you explain the reason why?

The `split()` function splits the text into words by the spaces that separate them and treats punctuation marks as part of the word. Consequently, the two occurrences of the word *day* are represented in the list as two different members: 'day' and 'day'.

In addition to punctuation marks, the English language presents us with another challenge regarding counting words in a text. Can you guess what it is?

Consider the following code. What do you expect the value of `num_w` to be?

```
num_w = words.count("it")
```

The word 'it' appears twice in the text. However, because the `count()` method distinguishes between uppercase and lowercase letters when it compares the value, it will not count the capitalized instance 'It'.

To overcome these challenges, we will need to "clean" the original text as part of a process called "text normalization", before we split it into a list and begin our data analysis.

Questions

10. *Hapax legomena (sg. hapax legomenon) refer to words which appear only once in a text. Write a function `GetHapax()` which receives a string value and returns the list of words in the string that appear only once. Test the function with different values.

3.7.2 Text normalization

The first step in this process will be to remove the punctuation marks from the string. For this we will use the `replace()` method that operates on strings. The method accepts two arguments – what we want to replace and what we want to replace it with – and returns as output the edited string. We can replace one character by another (e.g., all 'a' with 'b'), or, for our current purpose, replace the character that we want to delete with an empty string.

In the following code segment we execute the `replace()` method multiple times on the string variable `quote`. Each time we replace all instances of a particular punctuation mark in `quote` with the empty string and then re-assign the result to the same variable.

```
quote = quote.replace("?","")
quote = quote.replace(".","")
quote = quote.replace(",","")
quote = quote.replace("'","")
```

The value of quote after running this code segment will be as follows:

```
What day is it asked Winnie the Pooh It is today
squeaked Piglet My favorite day said Pooh
```

The next step will be to convert all uppercase letters into lowercase. This will be done with the `lower()` method, which applies to string variables and returns string values. The following code assigns the quote variable its value after all uppercase letters were converted to lowercase. A similar method, `upper()`, coverts lowercase letters to uppercase.

```
quote = quote.lower()
```

```
what day is it asked winnie the pooh it is today
squeaked piglet my favorite day said pooh
```

Now, after the text normalization process was completed, the string is free of punctuation marks and all the letters in it are lowercase, we can convert the string to a list of words and begin analyzing the data.

```
words = quote.split()
```

```
['what', 'day', 'is', 'it', 'asked', 'winnie', 'the',
'pooh', 'it', 'is', 'today', 'squeaked', 'piglet',
'my', 'favorite', 'day', 'said', 'pooh']
```

However, before we move on to the next step, we will take another look at what we have done so far.

In the process of "cleaning" the text we replaced each of the punctuation marks with an empty string, thus effectively deleting them. We did this using four replace commands, where in each of them we replaced a different punctuation mark. This procedure achieved our immediate requirements, but as computer programmers our goal is to design general solutions, which are not tailored to one particular case.

Can you think of an example of text for which this process would not be satisfactory?

The four punctuation marks that appear in our Winnie the Pooh excerpt are only a subset of the entire ranges of symbols that appear in texts and that should be removed. We will now consider a more general solution that handles all possible symbols.

Let us first define a string variable, punctuation, which includes the non-alphabetic characters that we would like to remove from the text.

```
punctuation = "!#$%&'()*+,-./:;<=>?@[\]^_`{|}~"
```

Now, we can simply iterate through the string using a for loop, and within the loop, in each iteration, call the replace method with the current character.

```
for x in punctuation:
    quote = quote.replace(x,"")
```

The use of a string of characters that are intended for deletion allows us to define them in one centralized location. In addition, while our original solution required us to write a `replace()` command for each character to be deleted, in the revised solution, the function appears once inside a loop, and the number of iterations matches the number of characters defined in the string.

Questions

11. *When we analyze the vocabulary of a particular text, we often want to focus on meaningful content words and disregard highly frequent function words (see discussion on concordance creation in Chapter 2). Write a program that takes a text, normalizes it by removing punctuation marks and changing all characters to lowercase, and then creates list which only includes the content words that appear in the original text (i.e., words that are not one of the following: 'a', 'of', 'on', 'for', 'with', 'the', 'at', 'from', 'in', 'to').

3.8 Input and output

3.8.1 Receiving input from the user

So far, the input in the programs we wrote has been part of the program itself. For example, when we converted a text into a list of words, we first defined a variable to store the text, and then performed operations on it.

```
quote = "'What day is it?', asked Winnie the Pooh.
'It is today,' squeaked Piglet. 'My favorite day,'
said Pooh."
words = quote.split()
```

In some cases, our program's output took the form of displaying content on the screen, achieved through the `print()` command.

Another way to provide input to a program is through interaction with the user via a dialogue. For us, as experienced computer users, this type of interaction with computer programs is very intuitive. Think of all the cases where you are asked to enter your username and password. Interactive computer programs that require input from users are designed to respond to any input they receive (including illegal inputs, such as an incorrect password). Naturally, as users, we are not exposed to the code and are not required to know how to program in order to use the software. The separation between the program and the input is a kind of modularity, or "separation of powers".

The function with which we will prompt the user for input is called `input()`. This function displays on the screen the text, which is specified as its argument, accompanied by a text box where the user can input a response. The user's response is returned by the function as a string.

```
quote = input("Input a sentence.")
words = quote.split()
```

Can you think of situations where this method of obtaining input is not appropriate?

One significant disadvantage of this method is that it requires the user to type the input. Later in the chapter we will want to examine the vocabulary of two short stories, and of course we will not want to type the full text. We will also prefer not to copy the entire text of a short story into a variable in our code, as we did for the short *Winnie the Pooh* excerpt.

To address this challenge, we will learn a third method: reading input from a file. In this method, instead of typing the string directly into the program or through a dialog box, we will save the text in a file. Our program will read the contents of the file into a string type variable and will subsequently perform operations on this string as we did when the text was typed directly in the code or by the user. If we want to run the program with different input, we simply update the contents of the file or read a different file.

3.8.2 Working with files

The files we will work with here are text files, whose extension is txt. Unlike Word files, the text in text files is "minimalist" and contains only readable unformatted characters.

In order to read or write to a file, we first need to open it by calling the open() function.

The function is called with two arguments:

- A string (or string type variable) specifying the full name of the file (i.e., a name that includes the path to access it). For example: "C:\work\python\somefile.txt" or "/content/gdrive/MyDrive/work/python/somefile.txt".
- The file access mode: reading ("r") or writing ("w") (or appending, which we leave for now).

   In the following code segments, we will just include the file name. However, when using Google Colab, the files need to be uploaded to the drive, and the following code needs to be executed in order to connect Google drive to the Colab environment.

```
from google.colab import drive
drive.mount("/content/gdrive", force_remount=True)
```

The open() function returns a file object type, which is then used to access the file and perform operations on it. If the file is not found, Python will return an error message.

```
f = open("somefile.txt", "r")
f = open("somefile.txt", "w")
```

The act of opening a file, whether for writing or reading, locks it, and does not allow another program to perform operations on it. Therefore, after we finish working, we release it using the close() method.

```
f.close()
```

3.8.3 Reading and writing

The file read and write operations are performed using the read() and write() methods that operate on the variable representing the file handle (f in the example above). We will illustrate these operations by using a text file called 'still-i-rise. txt', which contains the poem "Still I Rise" by Maya Angelou (the file can be found at https://zefsegal.com/computational-literacy-for-the-humanities).

Reading a text file

In the following code segment, the read() method is called after the file is opened. This method returns the text from the file, which is then assigned to the variable text. Finally, the file is closed using the close() method .

```
input_file = open("still-i-rise.txt", "r")
text = input_file.read()
input_file.close()
```

To observe the read text, we can print the value of text. Let us compare the output of the print() function with the actual value of the variable text, obtained by simply inputting the name of the variable as a Python command.

```
print(text)
```

```
Still I Rise
BY MAYA ANGELOU
You may write me down in history
With your bitter, twisted lies,
You may trod me in the very dirt
But still, like dust, I'll rise.
```

```
Text
```

```
Still I Rise\nBY MAYA ANGELOU\nYou may write me down
in history\nWith your bitter, twisted lies,\nYou
may trod me in the very dirt\nBut still, like dust,
I'll rise.
```

The output of `text` is the actual content of the text file. Where lines break appear in the output of `print()`, we find the special character \n. This character is a kind of command to the software, to begin a new line. When we use the `print()` command, Python executes this command and the text is displayed divided into lines.

Let's consider another method for reading a file.

Instead of reading all the text contained in the file into one variable using the `read()` method, we can read the file using a `for` loop that goes through the entire text line by line. For example, the code in the following code box reads each line into the `line` variable and prints only the lines in which there is a question mark.

```
input_file = open("still-i-rise.txt","r")
for line in input_file:
    if "?" in line:
        print(line)
input_file.close()
```

Writing to a text file

Now, instead of printing the output to the screen we will write it to a file. To do this, we will use the open() function with the "full name" of the file as the first argument. The second argument determines the mode, or in other words, whether the opening operation will overwrite previous contents, if they exist, or keep the contents and append the new contents to the end.

In the following code, open() is called with w as the second argument. In this case, the file does not need to exist. If it does, the opening operation will overwrite it and create a new, empty file in its place. The method write() is executed on the file object returned by open(). Its argument is the text that we wish to write in the file. In this case it is the string "Python". Similarly to read(), the last step is closing the file.

```
output_file = open("practice.txt","w")
output_file.write("Python")
output_file.close()
```

If we want to append another line to 'practice.txt', we do so with "a" as the second argument. The following code adds a line break to the end of the file and then writes in the new line "programming language". Without the line break the new string would appear adjacent to the previous one.

```
output_file = open("practice.txt","a")
output_file.write("\n" + "Programming language")
output_file.close()
```

Reading and writing

Let us now combine the two operations: open "angelou.txt" and copy all lines with question marks into a file called "questions.txt", instead of printing them on the screen.

```
input_file = open("still-i-rise.txt","r")
output_file = open("questions.txt","w")
for line in input_file:
    if "?" in line:
        output_file.write(line + "\n")
input_file.close()
output_file.close()
```

Note that we defined two different file object variables, one for each file: input_ file and output_file.

To avoid the need to remember to close the file after use, Python offers a command with open() which opens the file, and closes it automatically after all the operations described in the indented code below the command are completed.

```python
with open("questions.txt","w") as output_file:
    with open("still-i-rise.txt","r") as input_file:
        for line in input_file:
            if "?" in line:
                output_file.write(line)
```

3.9 Text analysis

We will now combine the knowledge we have gained so far in programming to analyze two types of texts. First, we will analyze a dialogue taken from the movie "Some Like It Hot" (1959) directed by Billy Wilder. The dialogue takes place during the meeting between Sugar Kane (Marilyn Monroe's character) and Junior (Tony Curtis's character).[1] Next, we will compare the literary styles of two short stories, "The Haunted Mind" by Nathaniel Hawthorne and "Two Kinds" by Amy Tan.

3.9.1 Some like it hot

The file which contains the dialogue from the movie *Some Like It Hot* (Wilder 1959) is called "some_like_it_hot.txt" (the file can be found at https://zefsegal. com/computational-literacy-for-the-humanities/). Each line in the file begins with the name of the speaker followed by a colon.

```
SUGAR: Haven't I seen you somewhere before?
JOE: Not very likely.
SUGAR: Are you staying at the hotel?
JOE: Not at all.
. . .
```

Our program will compare the number of words spoken by Sugar and Joe and then write the results in a new file, 'count.txt'.

Before we present the algorithm and corresponding code, you may wish to try to write them yourself.

```
# 1. Open the input file for reading
input_file = open("some_like_it_hot.txt","r")

# 2. Set the counters "sugar" and "joe" to 0.
sugar = 0
joe = 0

# 3. Loop iterates over lines in the input file
# Read the line into a string variable
for line in input_file:
  # Split the string into a list of words
  words = line.split()
  # Calculate number of words in line
  num_words = len(words)-1
  # If the first word is "SUGAR:"
  if words[0] == "SUGAR:":
      # increase "sugar" counter
      sugar = sugar + num_words
  else:
      # increase "joe" counter
      joe = joe + num_words
# 4. Close the input file
input_file.close()

# 5. Open the output file for writing
output_file = open("count.txt","w")

# 6. Write the results in two separate lines
output_file.write("SUGAR, " + str(sugar) + "\n")
output_file.write("JOE, " + str(joe) + "\n")

# 7. Close the output file
output_file.close()
```

The contents of the output file are:

```
SUGAR, 243
JOE, 257
```

3.9.2 *"The Haunted Mind" and "Two Kinds"*

We will conclude this unit with an exercise in which compare two short stories (the files can be found at https://zefsegal.com/computational-literacy-for-the-humanities/):

• "The Haunted Mind" by Nathaniel Hawthorne (1837)
• "Two Kinds" by Amy Tan (1989)

We will collect the following statistics:

Type and tokens: In text analysis, it's important to distinguish between **types** and **tokens**. Tokens refer to individual occurrences of words in a text, meaning that every time a word appears, it counts as a token, even if it's repeated. On the other hand, types represent unique words, so each word is counted only once, no matter how many times it appears. For example, in the sentence "The cat sat on the mat", there are six tokens but only five types, because "the" appears twice.

The split() method, which turns a string into a list of words, creates a list of tokens. To calculate the number of types we need to reduce the list of tokens by eliminating duplicate occurrences. An easy way to do this is to turn the list into a set using the set() function.

```
types = set(tokens)
```

The len() function can be used to calculate the number of items in both lists and sets.

Type–token ratio (TTR): TTR is a measure used in text analysis as an indication of the lexical richness of a text. It is calculated by dividing the number of types by the number of tokens in the text. Given two texts of equal length (i.e., the same number of tokens), the richer the vocabulary, the higher the number of types, and the higher the ratio. The type–token ratio is equal to 1 when the number of tokens is equal to the number of types, or, in other words, each word is repeated only once in the text.

Number of sentences: To count the number of sentences in the text we will apply the method count() to the raw text (the text before we normalized it and deleted all punctuation marks) and count the number of periods. Note that this is an approximation, since not every sentence ends with a period and not every period is necessarily an end of a sentence.

Average sentence length: The number of tokens divided by the number of sentences.

Number of quotation marks: Quotation marks are an indication of dialogs. To count them we will apply the method `count()` to the raw text. Quotation marks are special characters, so to search for them we must use a backslash `character \"`. This, too, is an approximation: there could be other uses of quotation marks in the text.

Questions

12. Analyze the two stories:
    a. *Read the text from the file into a string variable
    b. *Normalize the text (punctuation, lowercase)
    c. *Split the text into a list of tokens and a set of types
    d. *Calculate: number of tokens and types, type–token ratio, number of sentences, average sentence length, number of quotes.
    e. *Write the results into a text file
13. What do the results tell you about the difference between the two stories?

Even from the "dry" data one can see that these two stories have different styles. Amy Tan's story, "Two Kinds", is longer, but the sentences in it are much shorter compared to the sentences in Hawthorne's story, "The Haunted Mind". The difference in the number of quotation marks in the two stories hints that, unlike Hawthorne's story, Tan's story has many dialogues (70 quotes out of 319 sentences in total).

Another stylistic difference is reflected in the type–token ratio measure. In the story "Two Kinds", the relatively high type–token ratio (0.44) suggests a rich vocabulary, while in "The Haunted Mind", the lower ratio (0.28) suggest a more limited vocabulary. However, as we will see later in the course, there is great variance in the number of word repetitions in texts, so the simple type–token ratio we calculated does not tell the whole story.

Glossary

Append A method in Python that adds an item to the end of a list.

Assignment Setting a value to a variable using the `=` operator.

Boolean A data type with only two possible values: `true` and `false`.

Calling a function The process of executing a function by specifying its name and providing necessary arguments.

Concatenation Combining two strings into one using the `+` operator.

For loop A loop that iterates over a sequence (e.g., list, string) and executes code for each item in the sequence.

Function A block of reusable code designed to perform a specific task, which can accept inputs (called arguments) and return outputs.

If-Else Statement A conditional structure that executes different blocks of code based on whether a condition is true or false.

Index A position number used to reference an item within a sequence, such as a character in a string or an element in a list. Indexing starts at 0.

Input Function A function in Python used to get input from the user via the keyboard.

List A data structure that stores an ordered sequence of items of various types.

Logical operators Operators used to compare values, such as `>`, `<`, `==`, and `!=`.

Method functions defined for a specific type of variable. The call to the method is made only on variables of the appropriate type for that method.

Print Function A Python function used to display content, such as text or variable values, on the screen.

Remove A method in Python used to remove the first occurrence of a specific item from a list.

Reserved words Special words in a programming language that have specific meanings and cannot be used for other purposes (e.g., if, while, and).

String A sequence of characters, often used to represent text.

Syntax error An error that occurs when the code violates the syntax rules of the programming language.

Text normalization The process of preparing text for analysis by removing punctuation, converting to lowercase, etc.

Tokenization The process of splitting text into individual words or tokens.

Tokens The total number of words in a text.

Type–token ratio (TTR) A measure of lexical richness, calculated as the number of types divided by the number of tokens.

Types The number of different words in a text.

Variable A named location in a program that stores a value, such as a number, string, or list. Variables can be assigned new values during the execution of the program.

Note

1 See YouTube link: www.youtube.com/watch?v=udZgIsIKU30

References

Hawthorne, Nathaniel. 1837. *Twice-Told Tales*. The American Stationers' Company.
Tan, Amy. 1989. *The Joy Luck Club*. G. P. Putnam's Sons.
Wilder, Billy, director. 1959. Some Like it Hot. Mirisch Company. 2:01:00.

Solutions to selected questions

1. A program that prints the number of pages in the chapter.

```
first_page = 135
last_page = 136
print(last_page - first_page + 1)
```

2. Printing Boolean values.

```
print(8 == 8)
print(5 != 2)
print(3+4 == 2)
print(12 == 11)
```

3. Function questions.

```
def ratio (x, y):
    if y != 0:
        return x/y
    else:
        print("You cannot divide by zero!")
        return None
print(ratio(8,2))
print(ratio(8,0))
```

Note that None is used in Python to define "no value".

4. Checking the length of a string.

```
word = "apple"
if len(word) < 10:
    print(word + " has " + str(len(word)) + " characters")
else:
    print(word + " is too long")
```

5. Checking for spaces in a string.

```python
def HasSpace(text):
    if " " in text:
        return True
    else:
        return False

text = "This is a string"
if HasSpace(text):
    print("The text " + text + " contains a space.")
else:
    print("The text " + text + " does not contain a
    space.")
```

6. Counting the occurrence of a character in a string.

```python
num_a = 0
text = "abracadabra"
for x in text:
    if x == "a":
        num_a = num_a + 1
print(num_a)
```

7. Counting words in a string.

```python
num_words = 0
text = "Life is what happens when you're busy making
other plans."
for x in text:
    if x == " ":
        num_words = num_words + 1
print(num_words+1)
```

8. Counting list items in a text.

```
def IsLit(text):
    lit_words = ["poem", "novel", "author",
    "metaphor", "simile"]
    lit_count = 0
    for word in lit_words:
        if word in text:
            lit_count = lit_count + 1
    if lit_count > 2:
        return True
    else:
        return False

text = "Examples of literary devices are metaphors
and similes."
IsLit(text)
```

Note that although the text has literary content, only two `lit_words` items appear in it, so under the defined conditions it is does not count as literary.

9. Creating a list of long words.

```
words = ["poem", "novel", "author", "metaphor",
"simile"]
long_words = []

for w in words:
    if len(w) > 5:
        long_words.append(w)
```

10. Making a list of hapax legomena.

```
def GetHapax(text):
    text_words = quote.split()
    hapax = []
    for w in text_words:
        if text_words.count(w) == 1:
```

```
            hapax.append(w)
    return(hapax)

quote = "It was the best of times. It was the worst
of times."
print(GetHapax(quote))
```

11. Normalizing a text and creating a content word list.

```
rawtext = "It is a truth universally acknowledged,
that a single man in possession of a good fortune,
must be in want of a wife."

punctuation = "!#$%&'()*+,-./:;<=>?@[\]^_`{|}~"

stop_words = ["a", "of","on", "for", "with", "the",
"at", "from", "in", "to"]
content_words = []

# text preparation
cleantext = rawtext.lower()
for x in punctuation:
    cleantext = cleantext.replace(x,"")

text_words = cleantext.split()

# loop over words in the text
for w in text_words:
    # only append words not in stop_words list
    if w not in stop_words:
        content_words.append(w)  # only append
                                 # words not in
                                 # stop_words list
```

12. a.   Reading the text from the file into a string variable.

```
filename = "AmyTanTwoKinds.txt"
#filename = "HawthorneHauntedMind.txt"
input_file = open(filename,"r")
rawtext = input_file.read()
input_file.close()
```

b. Normalizing the text.

```
cleantext = rawtext.lower()
punctuation = "!#$%&'()*+,-./:;-?@[\]^_`{|}~"
for x in punctuation:
    cleantext = cleantext.replace(x,"")
```

c. Splitting the text to a list of tokens and a set of types.

```
tokens = cleantext.split()
types = set(tokens)
```

d. Calculating: number of tokens and types, type–token ratio, number of sentences, average sentence length, number of quotes.

```
num_tokens = len(tokens)
num_types = len(types)
type_token_ratio = num_types /num_tokens
num_sentences = rawtext.count(".")
avg_sentence_len = num_tokens/num_sentences
num_quotes = rawtext.count("\"")/2
```

e. Writing the results into a text file.

```
output_file = open("text_stats.txt", "w")
output_file.write("Tokens:    "  +    str(num_tokens)
+ "\n")
output_file.write("Types: " + str(num_types) + "\n")
output_file.write("Type-token  ratio:  "  +  str(type_
token_ratio) + "\n")
output_file.write("Num.   sentences:   "  +   str(num_
sentences) + "\n")
output_file.write("Avg. sentence length: " + str(avg_
sentence_len) + "\n")
output_file.write("Quoted texts: " + str(num_quotes)
+ "\n")
output_file.close()
```

The results:

The Haunted Mind

```
Tokens: 1791
Types: 787
Type-token ratio: 0.43941931881630375
Num. sentences: 52
Avg. sentence length: 34.44230769230769
Quoted texts: 1.0
```

Two Kinds

```
Tokens: 4547
Types: 1328
Type-token ratio: 0.29206069936221685
Num. sentences: 319
Avg. sentence length: 14.253918495297805
Quoted texts: 70.0
```

4 Reading charts

4.1 What is a chart?

A chart serves as a visual representation of information, enabling the consolidation and comparison of various data sets. It encompasses diverse types, including timelines, graphs, maps, networks, and tree diagrams. While some charts, such as geographical maps, have a long history, the widespread usage of charts is a relatively recent development, as discussed in the following section.

In contemporary times, charts pervade newspapers, websites, and social networks, providing essential information alongside written text. The term "infographic" is often used interchangeably with "chart" in the realm of visual communication. Over the years, charts have evolved from scientific tools used primarily by a select few to informational instruments for the general public for educational and marketing purposes. They are particularly prevalent in economic discourse, and it would be challenging to envision a business discussion without the inclusion of charts depicting profit and loss or changes over time.

In an era of abundant data, commonly referred to as Big Data, verbal presentation of information is becoming less practical, elevating the significance of charts as a crucial means of comprehending the broader context.

The reasons for incorporating charts into discourse are nearly self-evident:

- The graphical representation facilitates the presentation of a substantial amount of information concisely, aligning with the adage "A picture is worth a thousand words". This efficiency holds great value for knowledge producers and distributors.
- As a means of communication, charts allow for the unambiguous presentation of data, overcoming the inherent ambiguity of human language.
- As an interpretative tool, the strength of charts lies in the multi-dimensionality they convey. In contrast to verbal descriptions, which often present a singular perspective derived from the data, visual representations empower observers to synthesize numerous details independently, drawing their own conclusions. This openness fosters questions and encourages further discussions.

DOI: 10.4324/9781003502814-5

Thus, charts, in their various forms, have become the accepted mode of information transmission in the modern world. Their widespread usage, coupled with the rise of visual media, has redefined literacy in the 21st century. Modern literacy now extends beyond reading, writing, and vocabulary to include the ability to interpret visual signs with multiple meanings in different formats.

> So much data comes flying right at teens and young adults these days, and so many of them don't have the tools to know how to understand it, what questions to ask, or when it is just plain misleading in the way it is graphed. I'd like to see them be skeptical and be better consumers of information.
>
> (Gonchar and Katherine Schulten 2017)

In the words of Professor Judith Hunt from the Department of Mathematics at San Diego Miramar College, the prevalence of visual representation is not without flaws. The persuasive nature of charts, combined with the limited literacy of observers, often leads to the dissemination of false information (Huff 1954; Monmonier 1984). Critical analysis is essential when engaging with visual representations, considering the potential for misinformation inherent in many charts encountered daily.

4.1.1 A history of charts

Charts depicting quantitative information did not originate in the modern era; rudimentary examples can be traced back to ancient times. Some contend that the earliest recorded map is a clay tablet dating back to 3800 BC, discovered in 1930 in the city of Nuzi, Iraq. This tablet features a drawing documenting northern Mesopotamia. Others argue that an even earlier piece of evidence, a drawing from 6200 BC found in the excavations of Çatalhük in Turkey in 1963, serves as an indication of mapping. Despite differing opinions on the precise starting point, it is indisputable that the first maps emerged over 5,000 years ago.

As early as 3200 BC, Egyptian surveyors employed coordinates and visual representation to convey information about lands. In the Middle Ages, European astronomers described the movement of celestial bodies using circular lines on a grid that illustrated both celestial space and the dimension of time (Figure 4.1). By the 13th century, musical symbols had become a standardized notational language, thanks to the publication of Johannes de Garlandia's "De Mensurabili Musica" (concerning measured music, 1240), and Franco of Cologne's "Ars cantus mensurabilis" (The art of the measurable song, circa 1250). All of these examples represent types of charts conveying both quantitative and qualitative information through agreed-upon symbols and visual presentation.

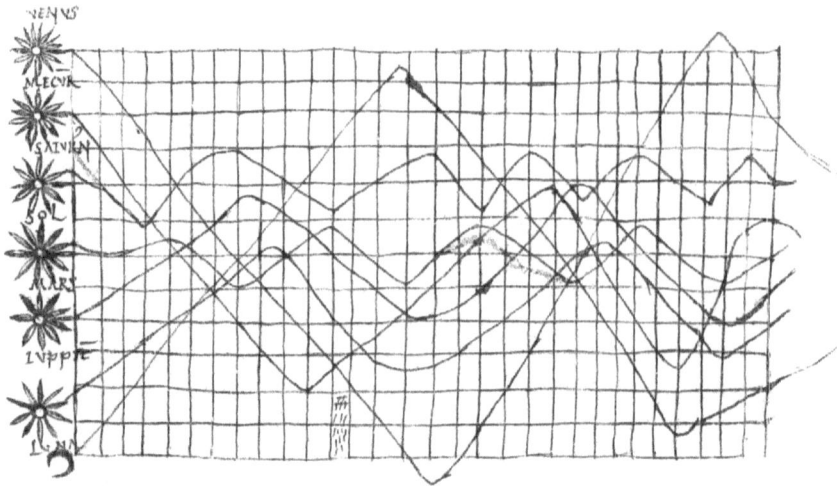

Figure 4.1 Movements of celestial bodies depicted as curved lines on a spatio-temporal grid. The diagram, authored by an anonymous individual, was featured in manuscripts of Macrobius' commentary on Cicero's work, "In Somnium Scipionis" (approximately 10–11th century).

Questions

1. Is it possible to understand Figure 4.1, which was created by an anonymous astronomer in the 10th century? What additional information could contribute to a better grasp of this chart?

While a few early diagrams have been presented thus far, the visual era of graphical information is widely considered to have commenced around the 19th century (Friendly and Wainer 2021).

Questions

2. Examine Figure 4.2 and answer the following questions:
   a. *What do the subheadings describe?
   b. *How does the chart depict the prevalence or scarcity of inventions over historical periods?
   c. *According to the diagram, when would you estimate the beginning of the "visual age", and what reasoning led to this determination?
   d. *What unresolved questions pertain to the development of the chart?

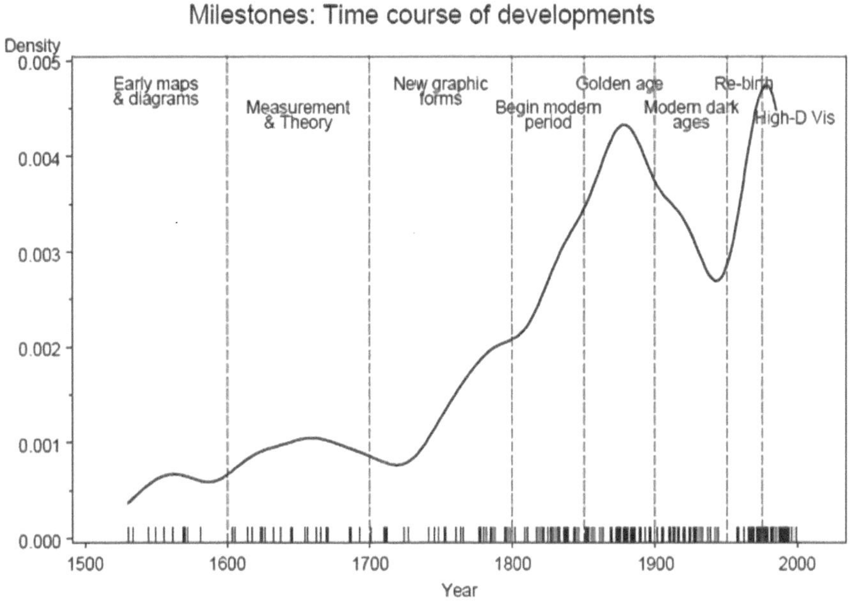

Figure 4.2 A timeline depicting the annual number of graphical innovations since 1500. The exact date of each innovation is represented by a single black line at the bottom of the image.

Source: Friendly and Wainer (2021).

We will not present a comprehensive review of the history of graphic representation but will focus on a brief description of its peak period – the 19th century – and the reasons for the development of charts during this period. During the first half of the 19th century, an exponential growth was evident in the development of new types of charts and their use. All modern types of graphic representations were invented during this period, including scatter plots, histograms, timelines, contours, pie charts, and graphs. Geographical mapping expanded from its focus on political entities (settlements, institutions, and borders) to the representation of information on diverse topics such as economy, society, health, and morality. This expansion required the invention of new graphic forms and signs.

It was not only about new inventions; there was also a significant surge in the widespread adoption of these visual representations. Visual analysis of natural, medical, and meteorological information saw a rising prevalence in scientific publications. Simultaneously, with the growing scientific use of visual displays, they appeared increasingly in textbooks and were incorporated into lesson plans within newly established state schools.

It is impossible to separate the expansion of the use of visual displays from their application within the administration and public service. The term "statistics" first appeared in a 1749 publication of Gottfried Achenwall (1719–1772), which dealt with the textual analysis of information about a country and its inhabitants. The word itself is derived from the Latin phrase "statisticum collegium" (council of state). Starting from the beginning of the 19th century, statistics became a field of thought specializing in collecting, summarizing, and analyzing information about the country's residents. Gradually, charts became the main way to present and analyze this information.

In the middle of the 19th century, following the processes of urbanization, industrialization, and the invention of the railroad, bureaus of statistics were established in many European countries. Their role was to assist in planning medical, economical, industrial, and transportation state policies. At the same time, mathematicians such as Carl Friedrich Gauss (1777–1855) and Pierre-Simon Laplace (1749–1827) helped develop a statistical theory, which was later extended to the social sciences by researchers such as André-Michel Guerry (1802–1866) and Adolphe Quetelet (1796–1874). These tools and methodologies made it possible to give meaning to substantial amounts of information, including the summarization of data using diagrams.

Florence Nightingale (1820–1910) is a good example of the social use of statistical tools. Following her work as a nurse in the Crimean War (1853–1856), during which she treated wounded British soldiers, she utilized statistics and charts to convince the British authorities that improving hygiene among the fighters could save many lives. Upon her return to Britain, especially in the years 1858–1859, Florence Nightingale, generally known as the founder of modern nursing, published a series of articles presenting the causes of the mortality among the soldiers of the British army verbally, quantitatively, and visually. A few of her articles were written as a response to critiques, who doubted her data and methods of analysis. Figure 4.3 shows one of Nightingale's most famous graphical innovations, known today as a "coxcomb".

Nightingale did not use the term "coxcomb" to describe the chart but rather the booklet of charts and tables that she added as an appendix to her report on the Crimean War. The use of this word referred to the likelihood that people would be more inclined to read the colorful and flamboyant appendix rather than the entire report.

The symbols of Figure 4.3 are explained in the diagram's legend.

The areas of the blue, red, and black wedges are each measured from the centre as the common vertex.

The blue wedges measured from the centre of the circle represent area for area the deaths from Preventable or Mitigable Zymotic diseases, the red wedges measured from the centre the deaths from wounds, and the black wedges measured from the centre the deaths from all other causes.

The black line across the red triangle in Nov. 1854 marks the boundary of the deaths from all other causes during the month.

In October 1854, and April 1855, the black area coincides with the red, in January and February 1856, the blue coincides with the black.

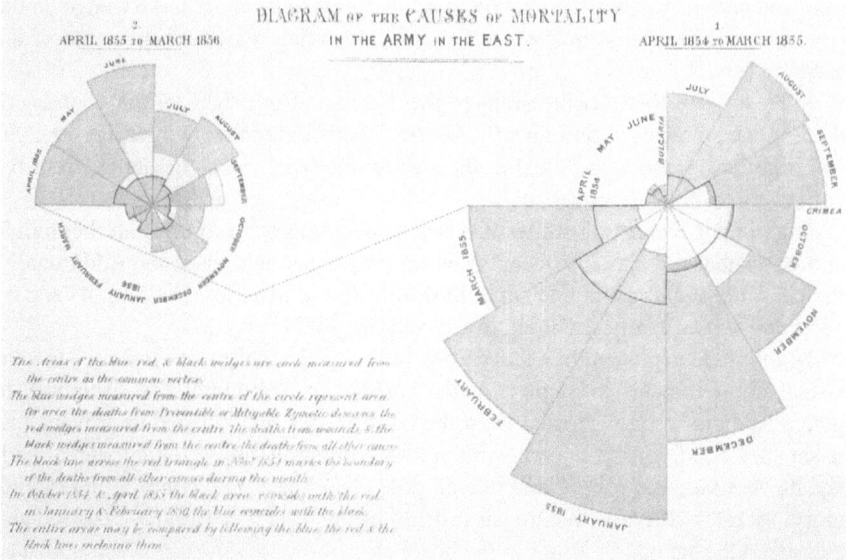

Figure 4.3 "The Mortality Factors of the British Army in the East", by Florence Nightingale. The diagram was published as part of an article written by Nightingale (1858) and submitted to Queen Victoria.

The entire areas may be compared by following the blue, the red, and the black lines enclosing them (the colored version of the diagram can be found at https://zefsegal.com/computational-literacy-for-the-humanities/).

Questions

3. Examine Figure 4.3 and answer the following questions:
   a. What do the two parts of the diagram represent and what do their visual differences imply? What is the purpose of the dashed line that connects both parts?
   b. What is the message of Nightingale's chart?

The heading above each part implies that the left diagram represents a later period (April 1855 to March 1856), while the right diagram depicts the preceding year (April 1854 to March 1855). The substantial difference in the sizes of the diagrams reflects a significant decrease in the number of deaths in the second year.

The dashed line is added to prevent misinterpretation. Without it, an observer might mistakenly perceive a cyclical pattern in the right diagram. Each wedge corresponds to a month, but the closure of the circle does not symbolize a cyclical progression of time; rather, it represents a fictitious connection between the beginning and end of the year. The line guides the observer from the large wedge on the right to its continuation in the left diagram, indicating the commencement of the second year.

Interestingly, Nightingale arranged the diagrams from right to left, contrary to the English language. Consequently, her readers would probably interpret the chart from the later period (left part) to the earlier one (right part), deviating from the correct chronological sequence.

Nightingale's chart illustrates that deaths resulting from injuries are negligible and consistent compared to deaths caused by preventable diseases. Additionally, the chart highlights a notable surge in deaths due to diseases, peaking in January 1855 and subsequently declining to almost none by January 1856.

Nightingale explained in a letter from 1857 that this shift was attributed to consistent improvements in hygiene at the hospital in Constantinople (Istanbul), to where soldiers were evacuated from the battlefield. Prior to these improvements, unsanitary conditions prevailed, with sewage contaminating drinking water, inadequate ventilation in rooms, and overall poor cleanliness – all contributing to high mortality rates. The positive transformation began after a committee assessed and addressed the sanitation issues in the hospital.

4.2 Types of charts

As established thus far, the number of potential diagrams is infinite, rendering attempts to categorize them into a finite set impractical. Nevertheless, there are several prevalent types that we will delineate in this section. The objectives of this description is to elucidate how to interpret charts as well as highlight their shortcomings.

4.2.1 Scatter plot

A scatter plot illustrates the relationship between two quantitative variables. Each data point is defined by two values, with the first value represented on the X-axis and the second on the Y-axis. This type of chart is effective for visually assessing the correlation or association between different variables which may be challenging to discern in other chart formats. Consequently, scatter plots are often used as an illustration of a linear correlation between two variables, a topic we will delve into further in Chapter 7.

Unlike many other charts invented during the 19th century for social analysis, the scatter plot was particularly effective in the natural sciences. Edward Tufte, a prominent researcher in graphic design, estimated that over 70% of charts in scientific publications are scatter plots. Therefore, it is not surprising that the first scatter plot was designed by a scientist.

The English scientist John Frederick William Herschel (1792–1871) explored various subjects, including the trajectories of double stars. For this research, he

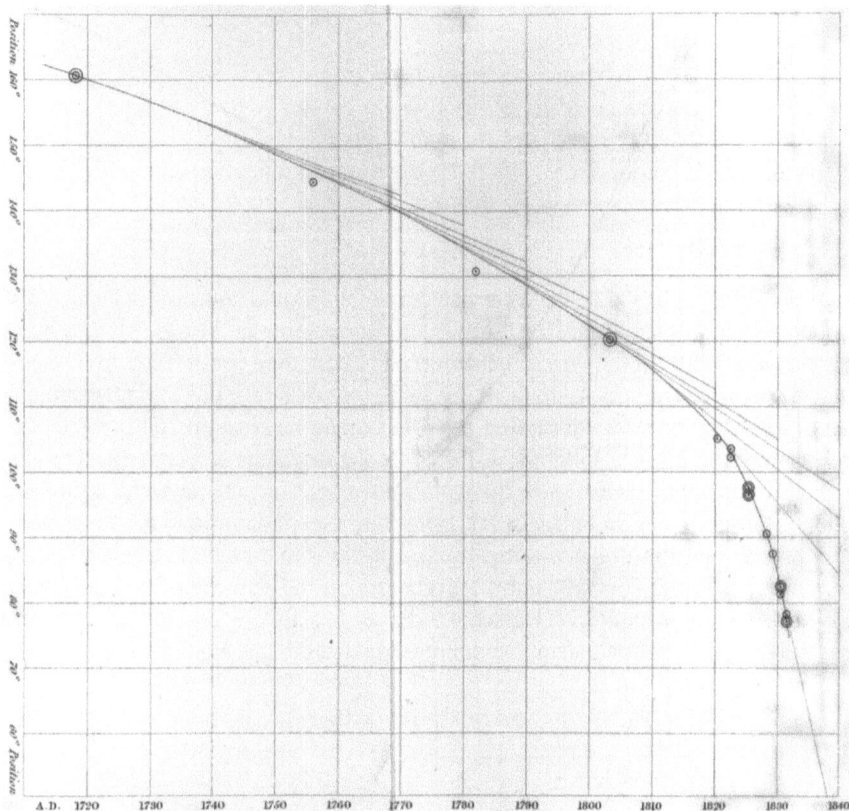

Figure 4.4 John F.W. Herschel's first scatter plot illustrating the observations of the double star gamma Virginis.

Source: Herschel (1833; published in Friendly and Wainer 2021).

devised a diagram presenting different observations of double stars using two coordinates (Herscel 1833). The Y-axis represented the angle at which the blue star was observed, while the X-axis denoted the year of the observation. Herschel unveiled his charts to the Royal Astronomical Society in London on January 13, 1832, and referenced them in a published article from 1833, but the charts themselves were considered lost until found again in 2006 (Figure 4.4).

There are two complementary ways to analyze a chart. The first is to observe it from a distance and discern patterns and clusters. One such pattern is the curve evident in Herschel's diagram (Figure 4.4), illustrating a relationship between the two variables. The second way is a closer examination of the data by concentrating on individual points and evaluating their relationship with their surroundings. The isolated observation from 1720 in Herschel's diagram is an interesting example.

Questions

4. On April 20, 1720, Jaques Cassini observed gamma Virginis at the Paris Observatory and noted its position angle as 140°. Herschel was aware of Cassini's observation and yet totally ignored it. Identify its hypothetical location in the graph. What are the disadvantages and advantages of retaining outlier points in the diagram?

Cassini's observation would have been an outlier and would thus disrupt Herschel's hypothesis (the curved line). The disadvantage of retaining outliers is that they may distract attention from the typical behavior, and often, they could result from inaccuracies in the observations. On the other hand, the advantage of keeping irregularities is that they open the discussion to criticism and reservations. This particular observation, for instance, holds significance from a research perspective as it might indicate a momentary change in the stars' movement or, alternatively, a flaw in Cassini's observations. In both cases, highlighting exceptions allows for the exploration of new questions related to the topic, questions that might not be addressed when focusing solely on the "norm". Furthermore, the purpose of a chart is not only to present a hypothesis (Herschel's curve) but also to provide the viewers access to the observational data, including those that deviate from the norm.

Questions

5. *The New York Times* published a peculiar type of scatter plot on November 7, 2017 in an article titled "The Words Men and Women Use When They Write About Love". It utilizes words extracted from personal columns written by readers over a four-year period in the newspaper's 'Modern Love' section. The journalists aimed to explore gendered language use and analyze the distinctions between columns eventually published in the newspaper and those left unpublished. In the chart, the points of the scatter plot were replaced by bubbles, thus providing another dimension of the data through the size of the bubbles. Accordingly, each point in a bubble graph is defined using four characteristics: a title, an X-coordinate, a Y-coordinate, and a size. The graph reappeared in the journal's website under the title "What's Going On in This Graph? Feb. 13, 2018" (www.nytimes.com/interactive/2018/02/08/learning/13WGOITGraphLN.html).

   View the graph, answer the following questions, and read the follow-up notes:
   a. *Describe the variables that appear in the chart.
   b. *Identify recognizable patterns in the visual image.
   c. *Examine the word 'love' and its surroundings.

i. Characterize the words that are identified as more feminine than 'love'.
ii. Is there any similarity between the prominent words that appear along the same vertical line as 'love'?
iii. Could you explain why a prominent word such as 'love' is identified as neutral in terms of its y-value?

4.2.2 Bar charts

A bar chart is one of the most prevalent graphical forms. Unlike a scatter plot, which illustrates a relationship between two quantitative variables, this type of chart depicts a relationship between a quantitative variable and a qualitative variable categorized into groups. Each category of the qualitative variable is represented by a column, with the height of the column indicating its corresponding quantitative value (see Figure 4.5). The taller the column, the greater the quantitative value associated with that category.

The bar chart made its debut in 1786 within the *Commercial and Political Atlas* of William Playfair (1759–1823), a prominent graphic innovator. The original chart (Figure 4.6) displays the imports and exports between Scotland and various parts of the world during a specific year. Playfair's creation of the bar chart is attributed

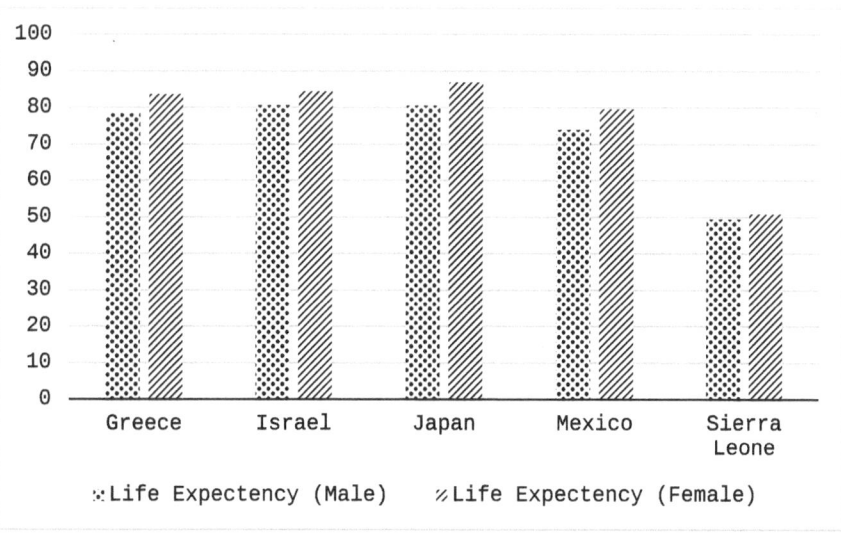

Figure 4.5 A bar chart presenting life expectancy in five distinct countries. The chart facilitates a straightforward comparison among the countries and between genders, enabling an immediate understanding of variations in life expectancy, indicating higher and lower values.

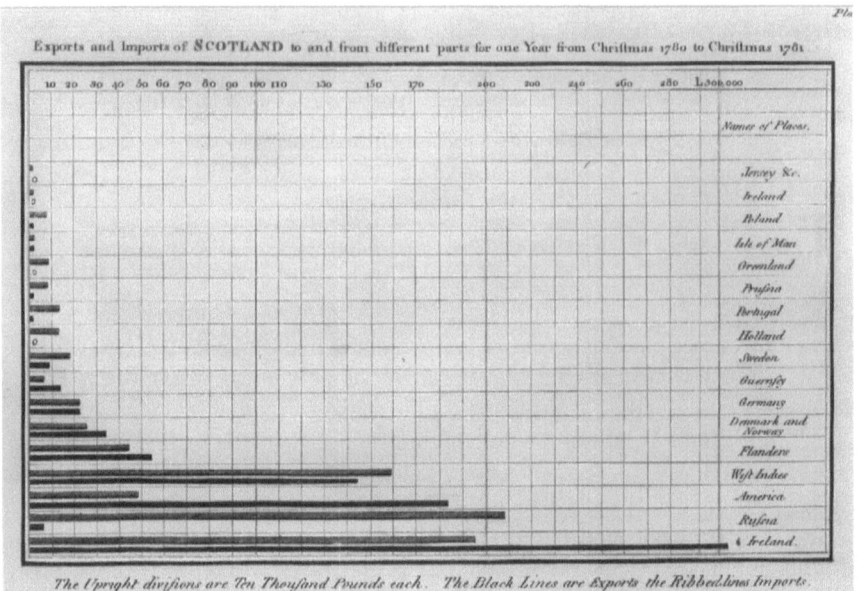

Figure 4.6 "Exports and imports of Scotland to and from different parts for one year, from Christmas 1780 to Christmas 1781" – the inaugural bar chart, published in Playfair (1786). Each trading partner is depicted by two columns, with the lower one representing exports and the upper one denoting imports. The values are measured in thousands of pounds for both exports and imports of goods.

to his lack of complete data. Wanting to showcase Scotland's foreign trade in his atlas, he limited his representation to the 17 trading partners for which he had comprehensive data.

"His bar chart", as noted by Beniger and Robyn (1978, 3), "was the first quantitative graphic form that did not locate data either in space ... or time". Similar ideas for representing information date back to the 14th century, with Nicola Oresme (1325–1382) exploring ways to depict speed relative to time. Playfair's innovation lay in presenting discrete and finite information rather than continuous data.

Bar charts are effective for illustrating comparisons between distinct categories. In Figure 4.7, for example, a bar chart is used to compare between literary genres according to their respective duration as prominent genres. The creator of this chart, Franco Moretti, utilized it as a foundation for his argument that a literary genre tends to have a relatively fixed lifespan, typically spanning 25 to 30 years. Furthermore, by utilizing accepted definitions of literary genres, Moretti used the diagram to argue that the exceptional case depicted on the right, representing "courtship novels", is not a singular genre but rather two distinct genres, each with a lifespan of 40 years.

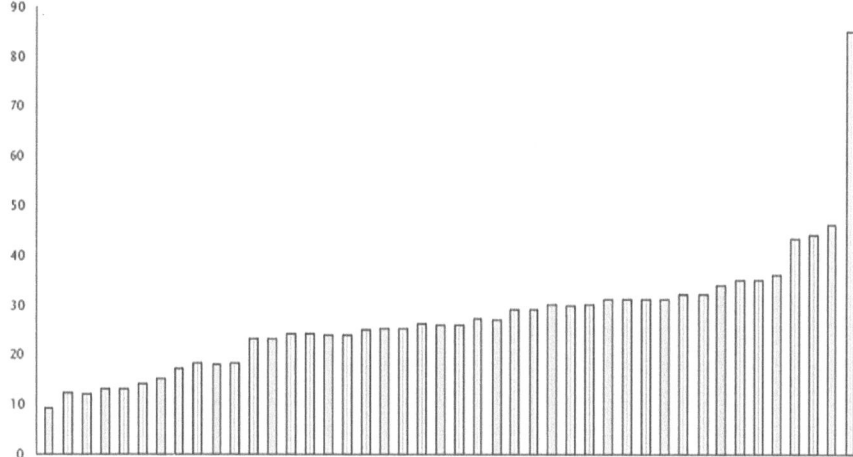

Figure 4.7 The number of years in which a literary genre was a prominent genre in the years 1740–1915. Each column corresponds to one of the 44 types acknowledged in research literature.

Source: Moretti (2003, 85).

Despite the historical significance and simplicity associated with bar charts, often being the first type of chart taught in schools, they exhibit a notable limitation. Bar charts are "visual tables" that fail to actively involve the observer in data analysis. The pre-established categorization dictates the interpretation, offering the observer little room for personal influence.

Questions

6. View the two bar charts that appeared in *The New York Times*' website under the title "What's Going On in This Graph? | Sept. 19, 2018" (www.nytimes.com/2018/09/18/learning/whats-going-on-in-this-graph-sept-19-2018.html), answer the following questions, and read the follow-up notes:
   a. What do you identify in these charts?
   b. What part of the charts would you like to investigate?
   c. Do these charts pose an argument?
   d. While the number of hurricanes making landfall has remained relatively consistent over the years, the cost of natural disasters has significantly increased. What conclusion can be drawn regarding the compatibility between the two charts?

The following response is not a direct answer to the questions posed, as these questions are intended for self-reflection. Instead, we will present the background of the charts, which you may have inferred from your examination.

The charts are featured as part of an article by *The New York Times* titled "The Places in the US Where Disaster Strikes Again and Again", published on May 24, 2018. From a formal standpoint, they combine elements of a bar chart and a timeline, as they depict varying quantitative values representing hurricanes along a timeline. The first chart illustrates the number of hurricanes making landfall in the United States from 1900 to 2017. The second chart depicts the total cost of natural disasters in the United States, specifically those causing an estimated billion dollars or more in damage, spanning the years 1980 to 2017.

Since the two charts describe different phenomena, drawing inferences from one chart to another without additional data may be misleading. The first diagram details hurricanes, while the second focuses on natural disasters, specifically those causing cumulative damage exceeding a billion dollars. It is important to note that these may represent distinct events.

However, the rising cost of natural disasters in the second diagram prompts questions about the impact of phenomena like global warming and world population growth on the cumulative cost of natural disasters.

4.2.3 Histograms

A histogram is akin to a bar chart but is specifically used to depict a continuous variable (rather than qualitative categories). In contrast to columns in a bar chart,

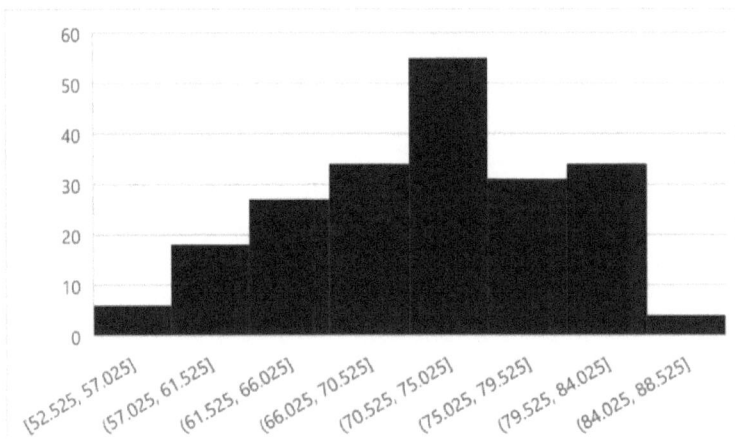

Figure 4.8 A histogram illustrating the life expectancy in countries worldwide. Each column corresponds to a range of life expectancy values, and the column height indicates the number of countries within that range. It is noteworthy that this histogram can be misleading as it does not consider the size of the countries. In this representation, a small country is given equal weight as a large country.

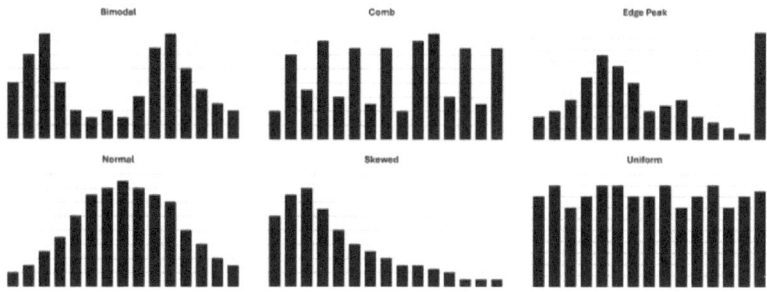

Figure 4.9 Six types of histograms.

where each column represents a distinct category, a column in a histogram signifies a continuous set on an axis. The width of the column corresponds to the size of the set it represents (see Figure 4.8). Unlike a bar chart, the placement of a column in a histogram is not arbitrary; it depends on the position of the set relative to the range of values.

For instance, if the range of values is 0 to 100, the column representing the range 0 to 10 will be positioned at the beginning of the chart, followed by the column for the range starting at 10. Unlike bar charts, where columns are distinctly separated, the columns in a histogram touch or nearly touch each other.

A histogram is often favored when identifying a distribution pattern proves challenging, and there are no clear categories for the data schema. Representing information across a continuous range simplifies the comparison of data in such cases.

Questions

7. What can we deduce from each of the six types, depicted in Figure 4.9, about the nature of the original data?

The **bimodal histogram** represents a variable with two distinct occurrences, suggesting the possibility of a cyclical phenomenon involving appearance and disappearance, resembling the pattern of car traffic which increases during rush hours.

In the **comb histogram**, the distribution alternates between high and low values. This histogram pattern may indicate challenges in processing or grouping the data into columns.

In the **edge peak histogram**, the variable is characterized by a peak at one end of the value range. In such a case, we would probably focus our attention on the cause of this statistical anomaly.

The **normal histogram** exhibits a normal distribution, a common pattern in many social phenomena. Most occurrences cluster around the center, gradually diminishing symmetrically as one moves away from the center.

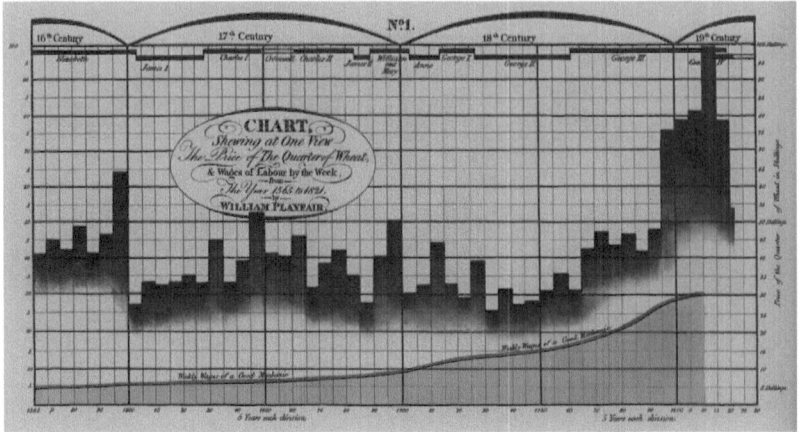

Figure 4.10 "A chart shewing at one view the price of the quarter of wheat & wages of labor by the week. The year 1565 to 1821".

Source: Playfair (1822).

The **skewed histogram** shows a distribution biased towards one end of the value range. Such a pattern often results from one end being bounded by a constant value, while the other end is infinite.

In the **uniform histogram**, there is an almost uniform distribution, indicating minimal differences in the number of repetitions of the phenomenon for various values within the range.

Unlike the bar chart, the histogram does not have a clearly defined inception date. The term first emerged in the lectures of the statistician Karl Pearson (1857–1936) around 1895, but examples of similar visual representations can be traced as far back as the early 19th century. In 1821, for instance, William Playfair published a chart (Figure 4.10) that concurrently displayed the prices of wheat in England from 1565 to 1821, the weekly wages of that period, and the names of the kings reigning during those years (Playfair 1822). The histogram represents wheat prices with black columns, the weekly wage is depicted as a continuous graph at the bottom, and the sequence of kings is presented at the top of the chart as a timeline.

Questions

8. Based on Figure 4.10:
    a. *Why did Playfair choose to present the price of wheat as a histogram, and not as a continuous graph, similar to the weekly salaries?
    b. *Identify the periods in which the life of wage earners was the most difficult, and those in which the life of wage earners was the most comfortable.

4.2.4 Pie charts

A pie chart is a circular diagram divided into segments which are not necessarily equal. The size of each segment corresponds to the relative size of the portion it represents within the entire group. This method is commonly used to compare distinct parts of a population. The advantage of the chart is that it provides an immediate visual representation of the relative sizes of the different segments.

The first presentation of this chart was in Playfair's book, *Statistical Breviary*, published in 1801 (Figure 4.11). Playfair's chart illustrates the relationships between the Asian, European, and African parts of the Ottoman Empire. The chart enables the identification and characterization of the relatively small portions of Europe and Africa within the Ottoman Empire, as of 1801.

While a pie chart is effective for illustrating relationships between parts and a whole, there are potential shortcomings that make many people hesitant to use it. The main limitations include the following:

First, in a pie chart, measurements are not exact and are based on estimated sizes. This makes it challenging to assess differences, especially when the values are relatively equal. For example, in Figure 4.12, three pie charts may appear similar, but as evidenced by the corresponding bar charts, they represent different distributions.

Second, a pie chart diminishes in value as the number of categories increases, as it becomes challenging to accurately estimate the size of sections, particularly when they are too small. For example, a pie chart illustrating the populations of all 50 states in the United States would be incomprehensible.

Third, a pie chart must add up to 100%. Its purpose is to demonstrate relationships between parts and the whole. If the sum of the different cases is incomplete, then

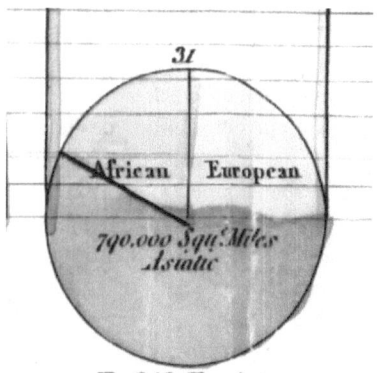

Figure 4.11 A pie chart from Playfair (1801). The chart depicts the relationships between the various parts of the Ottoman Empire across three continents – Asia, Europe, and Africa.

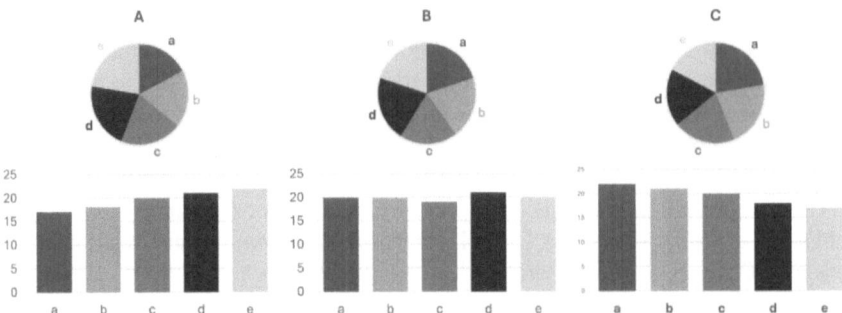

Figure 4.12 Three charts that appear almost identical, yet each represents a distinct
          distribution.

presenting the information in this way loses its significance, and it is more appropriate to use a bar diagram to convey the relationships.

4.2.5 Networks

Networks, or graphs, depict relationships between groups of entities. Each entity is represented by a vertex (or node), and the connections between the vertices are depicted by edges. The inception of the mathematical field concerning the study of networks is credited, as explained in Chapter 2, to Leonard Euler (1707–1783), and his solution to the Königsberg bridge problem. Since then, graph theory, as it is known in mathematics, and its counterpart in social sciences, network theory, have evolved into fields of research applied in various disciplines such as computer science, physics, biology, economics, and sociology.

A unique aspect of this field of thought is the amalgamation of quantitative theory and visual representation. In Chapters 8 and 9, the discussion on the topic of networks will be expanded, defining basic concepts and various mathematical properties for analyzing networks. However, in the current section, the focus is on understanding the diagram rather than its quantitative features.

It is crucial to remember that networks represent connections, so the spatial location has no significance, only the existence of an edge matters. Unless explicitly specified, the length, direction, or size of an edge does not alter anything. In Figure 2.8, the edge connecting A to D is not "longer", in terms of graph theory, than the edge from A to C. Similarly, the vertices B and C are not "down" or "up". These visual attributes hold no value in the network's meaning.

In 2015, Martin Grandjean created eleven charts illustrating the network structure of William Shakespeare's eleven tragedies (Figure 4.13). In these networks, two characters are connected if they appear in the same scene. Their size and color intensity are proportional to the number of their co-occurrences. Grandjean primary

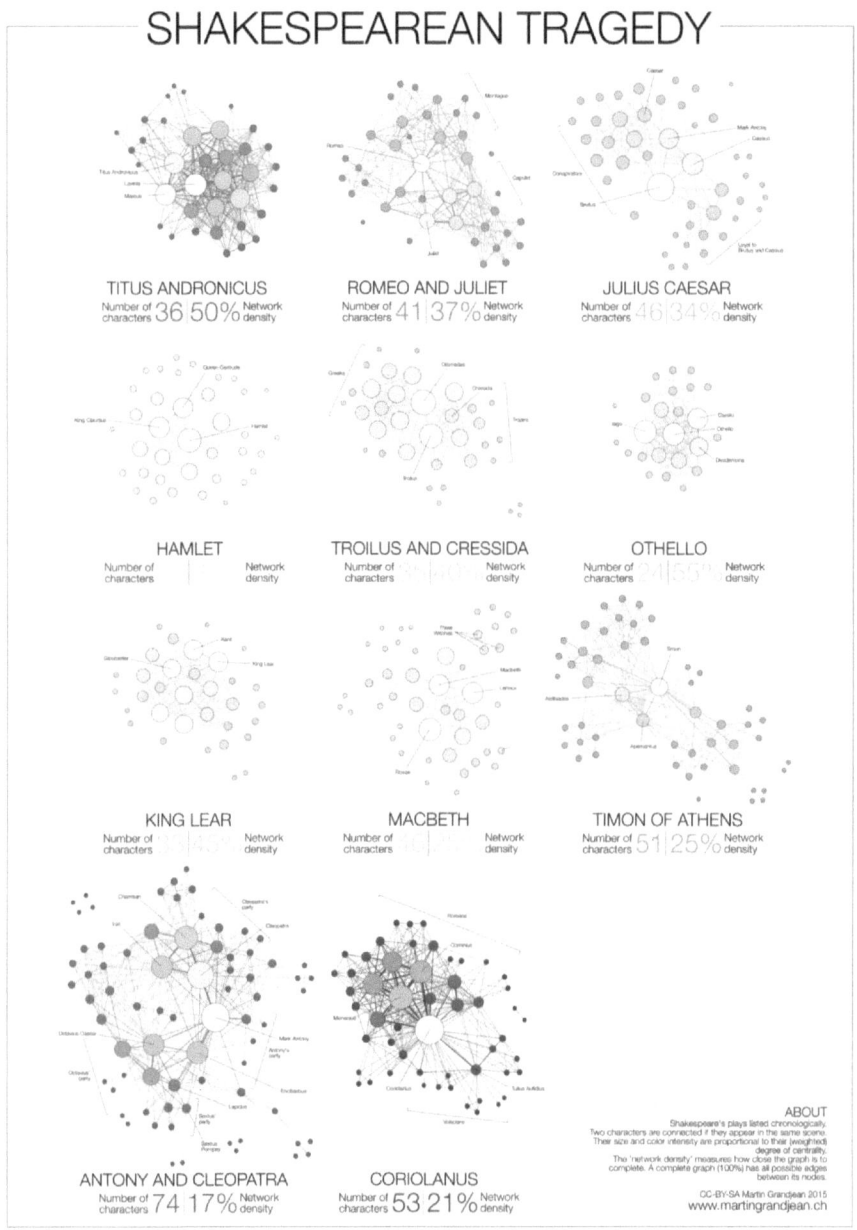

Figure 4.13 Networks representing Shakespeare's tragedies.

Source: Grandjean (2015).

question was: "Are Shakespeare's tragedies all structured in the same way? Are the characters rather isolated, grouped, all connected?"

Questions

9. Examine the networks illustrating Shakespeare's tragedies and answer the following questions:
    a. *In which tragedies do we observe a significant number of central characters, as per the network structure?
    b. *In which tragedies are there numerous extras?
    c. *In which tragedies is there a distinct division into subgroups?

Glossary

Bar Chart A graphical representation that compares different categories of a qualitative variable using columns whose heights represent corresponding quantitative values.

Chart A visual representation of data, commonly used to simplify complex information and enable comparisons. Examples include bar charts, pie charts, scatter plots, and histograms.

Coxcomb A graphical representation used by Florence Nightingale to present statistical data on mortality rates. It visually represents data in a circular format with wedges to denote different categories.

Histogram A type of chart used to represent the distribution of a continuous variable, where each bar represents a range of values and their frequency.

Infographic A visual representation of data or information, often combining text, images, and charts to communicate information quickly and clearly.

Line Chart A graph that uses lines to connect data points, showing changes in quantitative values over time or other continuous variables.

Network Diagram A visual representation of relationships between entities (represented as nodes) and their connections (represented as edges). Used in graph theory and social network analysis.

Outlier A data point that deviates significantly from the other observations in a dataset, potentially affecting the overall analysis.

Pie Chart A circular chart divided into segments, where each segment represents a proportion of the whole. It is used to compare parts of a population or dataset.

Scatter Plot A chart that displays points based on two quantitative variables, plotted along the X and Y axes. It is used to identify correlations or patterns between variables.

Statistical Graph A chart used to visualize data in a way that allows for analysis and interpretation of quantitative information, often used in scientific and economic discourse.

Timelines A type of chart that visually represents events in chronological order, often used to show historical developments or project progress.

Visual Literacy The ability to interpret and make meaning from information presented in the form of visual representations, such as charts and graphs.

References

Beniger, James R., and Dorothy L. Robyn. 1978. "Quantitative Graphics in Statistics: A Brief History". *The American Statistician* 32: 1–11.

Friendly, Michael, and Howard Wainer. 2021. *A History of Data Visualization and Graphic Communication.* https://friendly.github.io/HistDataVis/

Gonchar, Michael, and Katherine Schulten. 2017. "Announcing a New Monthly Feature: What's Going On in This Graph?". *New York Times*, September 6.

Grandjean, Martin. 2015. "Network Visualization: Mapping Shakespeare's Tragedies". www.martingrandjean.ch/network-visualization-shakespeare/

Herschel, John F. W. 1833. "On the Investigation of the Orbits of Revolving Double Stars: Being a Supplement to a Paper entitled 'Micrometrical Measures of 364 Double Stars'". *Memoirs of the Royal Astronomical Society* 5: 171–222.

Huff, Darrell. 1954. *How to Lie with Statistics*. Norton.

Monmonier, Mark. 1984. *How to Lie with Maps*. University of Chicago Press.

Moretti, Franco. 2003. "Graphs, Maps, Trees". *New Left Review* 24: 67–93.

Nightingale, Florence. 1858. "Notes on Matters Affecting the Health, Efficiency, and Hospital Administration of the British Army". Sent to Queen Victoria, October 11.

Playfair, William. 1786. *The Commercial and Political Atlas*. T. Burton.

Playfair, Willi am. 1801. *The Statistical Breviary*. T. Bensley.

Playfair, William. 1822. "Letter on our Agricultural Distresses, Their Causes and Remedies". Addressed to the Lords and Commons, London.

Solutions to selected questions

2. On "Time course of developments":
   a. The subheadings indicate the nature of innovations since 1500, and dashed lines demarcate different time periods.
   b. The presence or absence of innovations is reflected both by the height of the curved line (the graph) and the number of small lines at the bottom of the diagram.
   c. According to the diagram, there is a gradual increase in the number of graphic innovations starting from the mid-18th century. This trend peaks at the end of the 19th century, referred to as the "Golden Age", and experiences another peak at the end of the 20th century, labeled as a "Re-birth". However, caution must be exercised regarding the use of the term "era", as the chart does not indicate how these inventions affected reality. While, it illustrates the annual quantity of innovations, it does not show their application, leaving open the possibility that they remained esoteric and unused by anyone beyond their creators.

d. The diagram effectively highlights the quantity of graphical innovations and their periodization. However, it is unclear what qualifies as a "graphic innovation". How do we determine whether something is a noteworthy innovation, as opposed to a mere replication of previous designs? Additionally, not all innovations carry equal significance. The small black lines at the bottom represent each innovation equally, but it is clear that some changes have a more profound impact on reality than others. The headings at the top, along with the division into periods, are interpretive layers imposed by the author. They do not emerge directly from the data but result from a detailed content analysis of the graphical elements, rather than a numerical assessment of innovation quantities. As such, they should be viewed as part of the data interpretation rather than the data itself.

Moreover, the transition from individual black lines to a continuous curve raises questions about data abstraction. In Chapter 1, we discussed the distinction between finite sets, such as the number of innovations, and continuous sets, such as the curved line in the diagram. How was the set of black lines converted into an infinite set of points? Did this (imagined) abstraction lead to a potentially misleading interpretation of the data?

These questions do not invalidate the diagram but are essential considerations when interpreting its meaning.

5. On "The Words Men and Women Use":

a. The X-axis represents the extent to which a word is used more by men or women, with the information also conveyed through the color of the bubbles, which transitions between blue (for men) and pink (for women). The Y-axis measures the difference in a word's occurrence between published and unpublished essays. Although the chart does not explicitly state the meaning of the third variable, the size of the bubbles likely reflects the frequency of each word.

b. While it is difficult to provide a definitive answer, some patterns are notable:

   i. The central vertical axis, separating men from women, contains relatively few, spaced-out small bubbles. In contrast, the chart's corners feature larger bubbles, though these are relatively isolated. This pattern suggests that men and women use largely distinct vocabularies. Simultaneously, it implies that while the number of words used exclusively by one gender and appearing in only one medium (published or unpublished essays) is limited, their usage is relatively frequent.

   ii. Three dense clusters of points are observable: two located to the left and right of the center, and one in the upper-right corner. The first two clusters are composed of numerous small, dense bubbles, while the upper-right corner contains a significant number of large bubbles. This may suggest that women who publish essays and use feminine language tend to employ a shared vocabulary.

   iii. As observed by a reader of *The New York Times*: "women used more maternal and emotional words, whereas men had a broader range of words that were harder to find commonality in. As a result, the women had larger but fewer circles, while the men had smaller but more circles".

c. The word 'love' and its surroundings:

   i. Words that are identified as more feminine than 'love' are located to the right. They are mostly family-related terms such as 'mother', 'children', 'husband', 'daughter', 'sister', 'babies', marriage', or words expressing strong emotions like 'feel', 'cried', and 'pain'.

   ii. Words that appear along the same vertical axis as 'love', are more commonly used by women than men. 'love' appears relatively isolated on this axis, with a cluster of widely used words above it and a smaller group below. The words on this line tend to reflect positive emotions, including 'love', 'friends', 'home', 'wedding', 'marriage', 'heart', and 'beginning'.

   iii. The horizontal position of 'love' characterizes neutral words with no preference for published or unpublished essays. This suggests 'love' is used so frequently by women writers that it does not differentiate between the two mediums.

8. On "A chart shewing at one view the price of the quarter of wheat & wages of labor":

   a. Although the chart does not explicitly explain why Playfair presents the price of wheat as a histogram, it is likely that this choice highlights the fluctuations in wheat prices, which vary based on external factors like weather and foreign trade. The histogram allows for a more detailed representation of these changes. In contrast, wages tend to be more stable and continuous, which is better represented by a curve.

   b. According to the diagram, the periods with the largest gap between workers' wages and the price of wheat were at the end of the 16th century and the early 19th century. During these times, wheat prices increased while wages stagnated, likely leading to economic hardship for workers, as household staples such as bread and flour became more expensive. In contrast, the beginning of the 17th century, and even more so the mid-18th century, saw wages rise while wheat prices remained stable or fell, suggesting an improvement in living conditions for wage earners.

9. On Shakespearean tragedy:

   a. The size of each vertex corresponds to the number of its co-occurrences, indicating that *Titus Andronicus*, *Troilus,* and *Cressida* exhibit the largest number of central vertices.

   b. *Timon of Athens* and *Antony and Cleopatra* feature the highest number of extras, influenced not only by the large cast of characters in these plays (as indicated on the left under each tragedy) but also by the relatively few central characters in each.

   c. This question presents a challenge because the identification of distinct subgroups depends more on the network's visual layout than its inherent characteristics. While it may be tempting to identify distinct subgroups in *Timon of Athens*, *Antony and Cleopatra*, and *Macbeth*, it's important to note that this perception is largely influenced by the diagram's spatial arrangement, which does not provide sufficient information about the underlying distribution of vertices. Therefore, it is difficult to discern whether the distances between vertices are meaningful or simply a result of the layout.

5 Creating charts

In Chapter 4, we learned about different types of charts, their histories, and critical approaches we should use when reading them. This chapter complements the previous chapter by focusing on the process of creating charts.

We will present you with a challenge – creating charts that reflect data while considering their limitations. Each subsequent section serves as a case study for using charts in various areas of the humanities. In each case, you will be required to collect data, create a relevant chart, and critically analyze the resulting diagram. We will be using Google Sheets for basic charts and analysis, and Python for the final case study. This chapter assumes a basic acquaintance with creating and editing spreadsheets. If you have never used spreadsheets, you may refer to the Google tutorial at the "Google Workspace Learning Center".

5.1 Case study: "Small data" in art history

Following the technological advancements of recent decades, "Big Data" is a common buzzword. The relatively low cost of storage and the substantial quantity of data from diverse sources have shifted the focus of information technology toward data accumulation and management. It is estimated that in 2020, approximately 60 zettabytes of data were transmitted over the internet, which is equivalent to 60.2^{70} bits. For historical comparison, if all the information stored on the internet were evenly distributed among all the people in the world, each individual would possess 14,000 times more information than the content of the Great Library of Alexandria. Constructed during antiquity, its founders aspired to realize an ambitious vision – collecting all the existing information in the Hellenistic world. At its peak, during the reign of Queen Cleopatra, it housed about 900,000 different manuscripts.

The challenges of dealing with big data lead to the continuous development of data storage and retrieval methods, standardization of data writing, and algorithms for data mining and analysis. All these have significant value for researchers in the humanities, dealing with fields that encompass vast amounts of information. However, research in the humanities is still relatively small in scale.

DOI: 10.4324/9781003502814-6

In the field of art history, for example, where the scale of information is more "small data", there is no need for sophisticated methods for storing or retrieving data, and usually, data mining is performed by closely reading the sources. Nevertheless, the role of scholars is not confined to collecting data; they must give meaning to this information. To achieve this, connections and relationships between different pieces of information must be established and this integration can be quite complex. Therefore, leveraging computational tools or charts to identify relationships and patterns can still be highly effective.

5.1.1 On the history of artistic genres

In 1987, Paul DiMaggio (1987, 441) wrote the following:

> Literally, a genre is a "kind" or "type" of art. The notion of genre presumes that some aggregation principle enables observers to sort cultural products into categories. Formalists treat genres as comprising works that share conventions of form or content. Art historians also define genres in terms of shared conventions, but focus as well on social relations among producers in identifying "schools" or "artistic movements" [...] Although students of popular culture and literary theorists of the "reader-response" school consider formal similarities, they acknowledge that genres are partially constituted by the audiences that support them.
>
> Efforts of humanists to define genre in terms of form or content similarities represent attempts to impose normative order on systems of classification that are socially constructed. [...] The challenge for the sociology of art is to understand the processes by which similarities are perceived and genres enacted.

DiMaggio puts forth a thesis in his article, asserting that the division into genres is more closely tied to economic consumption than to formal resemblances among artworks. He suggests that "genre classifications let consumers invest in specialized knowledge and permit artists to do their work" (DiMaggio 1987, 445). In general, "creating artistic genres requires substantial investments. Rarely do artists even collectively possess the necessary wherewithal to organize the genre's potential constituencies (both artists and audiences) into a self-conscious, institutionalized art world".

It is not surprising, therefore, that painters are often associated with a specific genre. For instance, Salvador Dali is associated with the Surrealist movement and Louise Élisabeth Vigée Le Brun with the Rococo. However, it is evident that, in reality, artists are not limited to a single genre, especially considering that the genre itself is an artificial category created by external factors forced upon the artists and the art in question.

It is therefore interesting to examine artists based on the diversity of their classifications: do they tend to adhere to a specific artistic school, or do they express themselves in multiple ways? The answer to this question is not binary and can range from "monostylistic" to "eclectic".

An online resource for data regarding arts and their styles is WikiArt, a collaborative encyclopedia for visual arts, including information on approximately 250,000 artworks and 3,000 artists, as of December 2020. The site provides biographical information on various artists, their connections with other artists, and their characteristic styles.

For example, examining the works of Paul Gauguin (1848–1903), a French painter known for his paintings from French Polynesia and Martinique and associated with the Post-Impressionist movement, reveals a diverse range of styles.

Search for Gauguin on the website and press "view all 517 artworks".

Each painting is accompanied by information such as the place and date of creation, its style, the period in the artist's life, the current location of the artwork, and its dimensions. Under "styles" we get a breakdown of Gauguin's work: 251 artifacts are characterized as Post-Impressionist, 153 as Impressionist, 104 as Cloisonnist, 6 as Japonist, 5 as Synthetic, one as Primitivist, and one as Symbolic.[1]

These styles are relatively similar, and it would not be a surprise to find a Post-Impressionist artist associated with these styles. But was that the norm? To explore this issue, we will search for all French artists associated with the Post-Impressionist movement. To do this, perform the following actions:

Click on "Advanced Search" at the bottom of the page and choose "Post-Impressionism" under art movements and "French" under nation.

The search identifies 44 different artists, ranging from Henry Matisse with 1,008 artworks to Henry Moret with seven. To compare artists with relatively similar bodies of work, let's focus on those who have at least 400 artworks in the dataset.

The selected artists are:

- Henri Martin – 408 works.
- Henri Matisse – 1,007 works.
- Paul Cezanne – 588 works.
- Paul Gauguin – 517 works.

Create a table in Google Sheets and chart the style breakdown for each of these five artists.

Among the different options for representing these data, pie charts, in spite of their limitations, are likely the best for an immediate representation of the stylistic

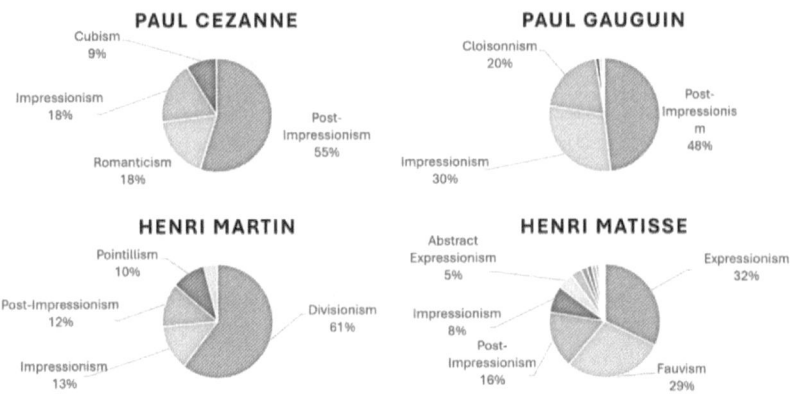

Figure 5.1 Pie charts reflecting the stylistic diversity of each artist.

diversity, if any, and the relative proportions of the prominent styles. Furthermore, they enable an easy comparison between artists.

Questions

1. Consider Figure 5.1:
    a. *Who is the most exceptional artist in terms of their diversity? Who are the two most similar among the five?
    b. *Examine the biographies of the four artists. Can you suggest a relation between the number of associated styles and their years of professional activity?
2. Create pie charts for all the Post-Impressionist artists who painted over 100 artworks, based on the information on WikiArt. Then, arrange the charts according to the birth year of the artists. (The image is available at https://zefsegal.com/computational-literacy-for-the-humanities/)
3. Explain the emerging pattern and identify outliers that deviate from the historical logic you identified.

Out of the 16 charts, there are five (Gustave Loiseau, Édouard Cortès, Maurice Utrillo, Moise Kisling Gustave Loiseau, and Paul Jacoulet) that depict artists whose works are characterized by only one style. Artists of this kind, whether they are monostylistic as the charts describe, or whether this determination arises from the catalogers' lack of attention to nuances, are not particularly useful in understanding the development of genres.

When examining the other charts, we can see a gradual development of genres in the later stages of the 19th century. Up until Henri Mattise, both in 1869, all artists have one dominant genre characterizing at least 45% of their work, reinforcing the idea that artists tend to maintain a consistent style. However, Mattise, along with Raoul Dufy and Andre Derain, born between 1869 and 1880, are no longer defined by a single genre, but by a variety of artistic styles. More broadly, as humanities scholars, when interpreting charts, we should avoid focusing solely on the most prominent features and brightest colors, as this can obscure important nuances and deviations. By examining the peripheral genres, we can see that their presence steadily increased in the last quarter of the 19th century.

"This period, covering the middle to the late 19th century and the first quarter of the 20th century is the stage in which the 'traditional' definition of both the artist as a social type, the artistic career and the production and dissemination of 'fine' cultural works […] comes to be defined and consolidated" (Lizardo 2008, 15). During this time, artistic innovation had become a market category, which lead to a proliferation of art movements and styles. In particular, Impressionists asserted their "controversial innovation" and influenced the rise of other contemporary avant-garde movements (Delacour and Leca 2017). As DiMaggio claims, the phenomenon identified in these pie charts is not just the development of artists and art but also the development of social and economic constructs. Interestingly, four of the five monostylistic artists appear toward the end of our chronological list of Post-Impressionist painters, a pattern that raises new research questions, though they will not be explored here.

5.1.2 On the history of artwork dimensions

Studies have shown that consumer culture accelerated during the 18th century. In 1714, Joseph Highmore, an English painter, recounted a visit to a London colleague:

> He hired a long garret ... where he painted cloths many feet in length ... and painted the whole at once, continuing the sky ... from one end to the other, and then several grounds etc., til the whole was one long landscape. This he cut up and sold by parcels as demanded ... and those who dealt in this way would go to his house and buy three or four, or any number of feet of landscapes.
>
> (Whitley 1928, 23)

Culture had been commodified. Paintings, books, and musical works passed through merchants and auction houses before reaching the hands of intended consumers. Technological advancements that reduced the cost of paper and rendered duplicating works of art more feasible, democratized access to art. As wealth disseminated among the middle class and market competition intensified, consumers diversified.

A study conducted by Carol Gibson Wood (2002), examining inventory lists of London houses from 1695 to 1715, reveals a notable shift in the art landscape. In many middle-class households, paintings began to adorn the walls, with approximately 45% of family homes in London possessing some form of artwork, be it paintings or prints. On average, each of these households displayed around 12 pictures. Given that London had about 25,000 households during that period, it can be inferred that the early 18th century saw around 135,000 pictures gracing the homes of the London middle class. This surge in buyers resulted in a significant decline in the proportional representation of the traditional aristocracy among art patrons, replaced by an increase in the participation of clergymen, art collectors, rural aristocrats, professionals, and aristocratic women. This diverse group displayed a lesser adherence to the artistic conventions of the past (McNally 2014, 21).

There are various approaches to examining the emerging audience that engaged in art consumption and to exploring the dynamic interaction between the culture of consumption and the inherent nature of art. One such method involves exploring the material artifact. An illustrative instance of this approach is the research conducted by Hege Roivainen et al. (2019) on 18th-century reading culture. In their study, these two researchers compiled data on the dimensions of over two and a half million books printed between 1500 and 1800, and they scrutinized the culture of reading by analyzing the evolving sizes of these books (Figure 5.2).

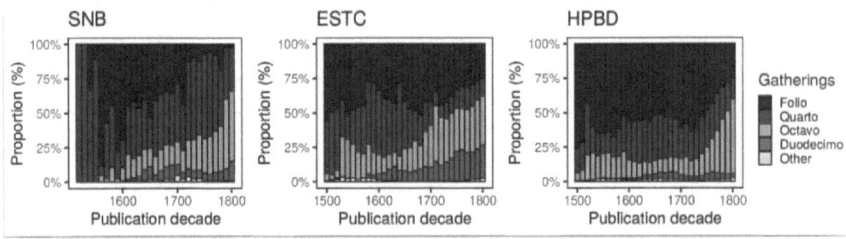

Figure 5.2 Shares of book formats according to print area in Sweden (SNB), Great Britain (ESTC), and the rest of Europe (HPBD). Print area stands for paper used for one copy of a book.

Source: Roivainen et al. (2019).

Questions

4. *Figure 5.2 depicts the changes in the size of books from the 16th to the 18th centuries. Is it possible to identify any trends during these 300 years?

According to the two researchers, these findings are directly linked to the shift in book-buying preferences, with readers increasingly favoring pocketbooks designed for private reading over larger books intended for public recitation.

These assertions align with descriptions of Paris by an unknown individual in the 1790s:

> Everyone, but women in particular, is carrying a book around in their pocket. People read while riding in carriages or talking walks; they read at the theatre during the interval, in cafes, even when bathing.
>
> (Wittmann 1999, 285)

The relationship between the size of a book and its cultural "importance" was also satirically depicted by British author and journalist, Joseph Addison (1712), at the outset of the 18th century.

> I have observed that the author of a folio, in all companies and conversations, sets himself above the author of a quarto, the author of a quarto above the author of an octavo, and so-on, by a gradual descent and subordination, to an author in twenty-fours ... In a word, authors are usually ranged in company after the same manner as their works are upon a shelf.

Addison's satire accentuates the significance attached to the size of cultural artifacts, underscoring the historical success of creators of "big" books.

5.1.3 Artwork size across artists and time

We will now attempt a similar exploration of artwork's sizes using the WikiArt database. To illustrate, let's consider the paintings of Jacques-Louis David (1748–1825), a prominent French painter associated with the neo-classical school. David is particularly recognized for his contributions to the Historical Painting style, which held a preeminent status in the early modern period. This style often featured depictions of grand, multi-participant scenes portraying historical, mythological, or epic events.

Ten paintings of David are labeled in the Wikiart.org website as "famous". We will limit our study to these paintings, while acknowledging that this methodology introduces potential risks (can you think what they are?). Paintings appear on the website with information that is relevant to our study: date and dimensions. We will retrieve this information for each of the ten paintings and store it in a Google Sheets table. To facilitate an appropriate comparison, we will use the area (height×width) as an indicator of the paintings' size.

Title	Year	Height	Width	Size (height x width)
Patroclus	1780	121.5	170.4	20703.6
...	...	...	...	...

Questions

5. Plotting the data:
   a. *Which type of diagram would be most suited for plotting the data?
   b. *Which columns would be included in such a diagram?

A scatter plot, such as Figure 5.3, would be the most suitable diagram to reflect the relation between the size of the artifacts and their chronology. However, assessing David's work in isolation, without comparing it to the works of other painters, is not sufficient for forming a comprehensive evaluation. For this reason, we will create a similar chart, yet include in it also works by the following French artists: Eugene Delacroix (1798–1863), Jean-Baptiste-Camille Corot (1798–1875), and Hilaire-Germain-Edgar De Gas (1834–1917), each artist representing a different artistic style and a slightly different historical context.

First, we will compile a table with the relevant data for each painting: the name of the artist, the painting's publication year, and the painting's size. Similarly to the previous chart, the X-axis will be the publication year and the Y-axis the size of the paintings. In order to distinguish between the different artists by using different colored dots we will need to associate the size of the painting with its respective artist. To do so, we will create a separate "size" column for each artist, enter the size of the painting only in the appropriate column, and "#N/A" in the other columns. The scatter plot will include data from all columns except the first column.

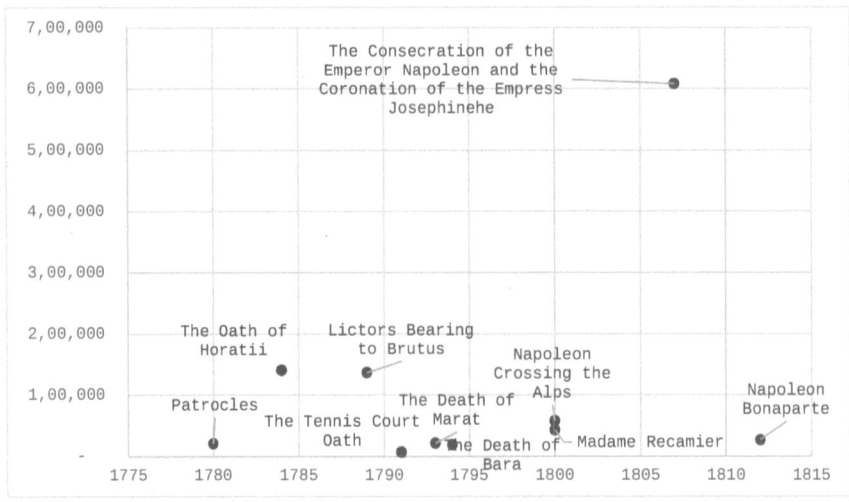

Figure 5.3 Date of publication versus painting size of Jacques-Louis David's most prominent paintings.

Artist	Year	David Size	Delacroix Size	Corot Size	Degas Size
David	1780	20703.6	#N/A	#N/A	#N/A
...	...	...			
Delacroix	1822	#N/A	45643.5	#N/A	#N/A
...	...		...		
Corot	1834	#N/A	#N/A	1218	#N/A
...	...			...	
Degas	1860	#N/A	#N/A	#N/A	50000
...	...				...

Questions

6. What methodological challenge arises from including the artwork *The Consecration of the Emperor Napoleon and the Coronation of the Empress Josephine* by David?

David's *The Consecration of the Emperor Napoleon* is 3 times larger than the second largest painting in this dataset. Therefore, it is an outlier that obscures the differences between the other paintings. To overcome the challenge, we will create a diagram that excludes *The Consecration*.

Questions

7. Consider the diagram in Figure 5.4:
   a. What historical pattern emerges from the diagram?
   b. What difficulties might arise if we attempt to draw conclusions from the diagram?
   c. Examine the works of David, Delacroix, Courbet, and Degas closely. In your opinion, what characterizes the content of the "large" paintings, and what characterizes the content of the "small" paintings? Furthermore, why is David's 1807 painting so notably large in its dimensions?

The diagram reflects a development similar to that described by Marianne and Ravouvain regarding the culture of reading: during the 19th century, artwork became smaller. One can only speculate that this is due to the adjustment of their sizes to consumer demands.

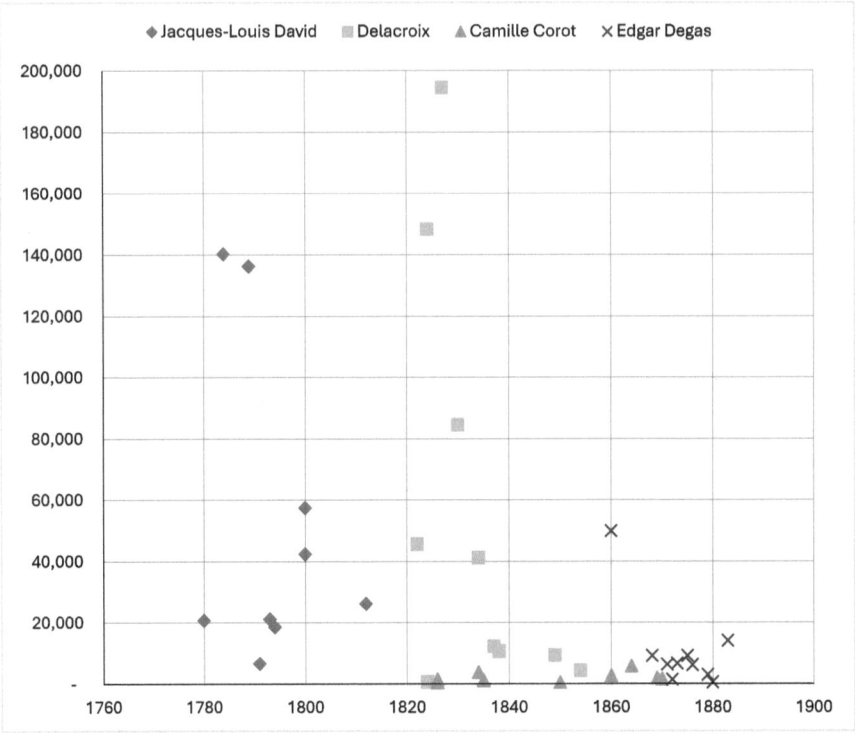

Figure 5.4 Date of publication versus painting size of the four artists' prominent paintings, excluding *The Consecration.*

The previous insight should be taken with a grain of salt since our sample is small. In the examined sample, there are 40 artworks from four different artists spanning various periods. It is challenging to argue that this sample is representative and faithfully reflects historical development. To strengthen our claim, we need to examine a larger, more representative sample.

Let's return to the *Wikiart* website and inspect the paintings themselves. It's hard not to notice the differences among the various creators. Jacques-Louis David painted visible, historical events full of pathos. Delacroix's paintings differ from David's but also depict historical or mythological events rich in detail. In contrast, Courbet and Degas' paintings are less detailed and closer to everyday life. One can only speculate about how *The Dance of the Nymphs* (Corot, 1850) would look in the hands of Delacroix: dancing nymphs would likely replace Courbet's large trees at the center of the painting.

The genre and size of paintings are often related. Historical and mythological paintings tend to be larger, fitting their visually dramatic subjects, while everyday paintings tend to be smaller, catering to a broader audience. It's no wonder that

David's painting depicting Napoleon Bonaparte's coronation is monumental. This work was commissioned by Napoleon himself and exhibited at the 1808 and 1810 Salon. Its monumental size was exceptional even by contemporary art.

Despite obvious reservations, due to sample size, the historical process we identify in the diagram reflects a change in consumer culture and demographics. In fact, the two main genres in 17th-century art, historical painting and portraiture took contrasting paths in the 18th century (Lippincott 2013). Historical painting struggled to maintain its existence due to a mismatch between the values it represented, its monumental scale, and the significant effort required for its creation, versus the new consumer audience seeking paintings related to their daily lives. Concurrently, the art of portraiture flourished, and its demand increased. Most portraits were small and uniform, and the drawn subject often paid for the right to sit for the artist. The equipment required for portrait painting was portable, allowing artists to work anywhere, whether in their personal studios or the client's home. The livelihood of portrait painters was more secure compared to painters of historical works.

Much like Roivainen's research on reading culture, our mini-research was able to bring to the fore a material change. However, much like any computational exploration in the humanities, the diagram and the data were only the starting points of a closer analysis of the content and reception of the paintings.

5.2 Case study: Exploring change over time in political history

Israeli parliamentary democracy, like other multi-party parliamentary democracies worldwide, relies on coalitions as a bridge between the legislative and executive branches. As early as 1933, 15 years before the establishment of the State of Israel, David Ben-Gurion articulated the essence of a coalition:

> The meaning of a coalition is not friendly bonds, but rather action and responsibility. The first condition for a coalition is a shared action plan. A coalition is not an end but a means. Only when there is a shared plan in which everyone can take part, it becomes a coalition. The second condition for a viable coalition is competent leadership, a group of individuals who not only represent their parties but also share mutual respect and trust, allowing for collaborative action. Without these two conditions, a coalition has no real value. It is merely a deception.
>
> (Quoted in Yaakobi 1980, 71)

The political reality significantly differed from the idealistic sentiment conveyed by Ben-Gurion's words. In the very first coalition negotiations in 1948, a conflict emerged between the partisan expectations of coalition members and Ben-Gurion's vision for coalition unity.

> Naturally, Ben-Gurion and his party [Mapai], having secured the majority status in the government, aimed to diminish the demands of coalition parties and broaden the collective constraints imposed on them by participating in the

government. On the other hand, the other coalition parties, already part of the interim government and aspiring to be part of the first coalition government, sought to maintain the pre-independence tradition [of co-leadership]. This was driven by their distinct demands as well as their concerns about Mapai's potential dominance due to its plurality rule. [...] Ben-Gurion aimed to forge a new and centralized coalition tradition, even in the absence of an alternative to a coalition government. The [political] split in the First Knesset and Mapai's pivotal position within it provided Ben-Gurion with the opportunity to do so, albeit at the expense of frequent coalition crises and a deepening of cumulative inter-party tensions.

(Yanai 1987, 177–178)

Since the inception of the first government to the present day, no party in Israel has ever secured enough votes to independently form a government without involving other parties. Consequently, coalition crises, and occasionally coalition unity, have become an integral and perhaps necessary aspect of the history of the State of Israel.

5.2.1 Coalition stability I

The stability of a government in Israel is not assured, as evident from the fact that, in the first 72 years of the state, 35 governments have served. However, there is a lack of uniformity in the duration of government tenures. Alongside a government that served for only 3 months, such as the 16th government, there was also a government that served for 5 years, like the 34th government. This raises the question of whether the structure of the coalition, and particularly the depth of its parliamentary support, contributes to the stability of the government.

In Israel, unlike other countries with coalition governments such as Norway, Belgium, and Germany, there is no inclination to form a coalition with the minimum parliamentary majority. Instead, there is a tendency to seek a broad coalition, often forming "surplus" partnerships beyond what is strictly necessary for a majority in the Knesset. In fact, only three coalitions began their term as minimal coalitions: the coalition led by Menachem Begin in 1981, the coalition led by Yitzhak Rabin in 1992, and the coalition led by Naftali Bennett in 2021.

Questions

8. Our data includes the sequential number of the government, the year it began its term, the duration of its term in months, and the maximum number of parliamentary members supporting the coalition (data available at https://zefsegal.com/computational-literacy-for-the-humanities/). Which type of chart would best illustrate the relations between parliamentary support, term duration, and time? Experiment with different types.

A scatter plot where the X-axis represents the year of the government's inauguration is likely the ideal option, but line charts or bar charts can also be effective. In the following chart, parliamentary support and duration of the term are represented using two lines.

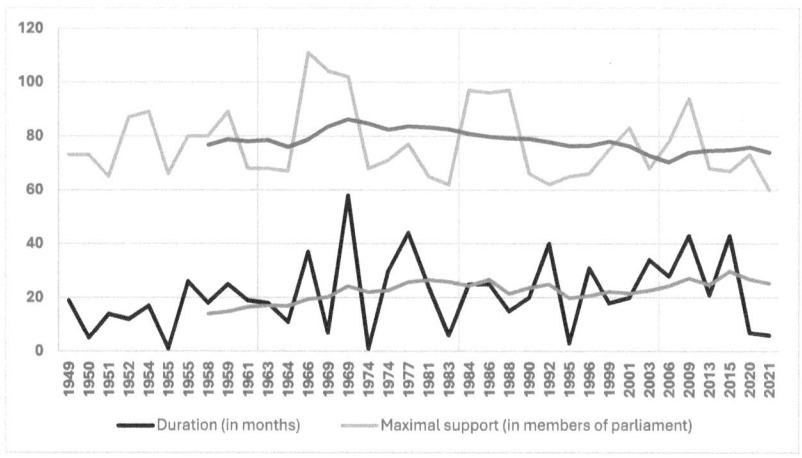

Figure 5.5 A timeline depicting the duration and parliamentary support for each Israeli government (1948–2022), as well as the corresponding trendlines.

Questions

9. *What advantage does this chart have over a bar chart when trying to simultaneously display both metrics?
10. Examine Figure 5.5. Is there a clear pattern emerging from the chart, and if not, why do you believe there is no such pattern?

A glance at the chart reveals two characteristic features of a typical Israeli government: a relatively short duration despite relying on a broad coalition. However, these observations do not necessitate a chart since they arise from the tabular data. An average government serves just 21.6 months (compared to a full duration of 48 months) and relies on 77 parliament members (out of a total of 120).

A chart is not a useful method to reveal averages but rather changes, distribution, and relationship between variables. In relation to these aspects, a clear pattern is not apparent at all. The chart displays governments with varying term durations that seem to have almost no correlation with the size of their coalitions. Additionally, there appears to be no correlation between the coalition's size, the government's duration, and the historical period in which the government reigned.

The conclusion drawn from this is that when attempting quantitative analyses on historical processes, particularly the chaotic nature of coalition structures, mathematical regularities may not apply.

While it may be difficult, perhaps even impossible, to characterize each point as part of a clear pattern, it's possible that a pattern can be identified if examining the change of averages over time. To do this, we will use a trendline – a line highlighting the general direction in which the data is moving over time.

Plotting trendlines

I.   Highlight one of the lines in the chart, then double-click on it. A "Format Data Series" window will open in the chart editor.

II.   Choose the "Trendline" option that appears in the settings.

III.   There are several types of trendlines based on the nature of the data. We will select "Moving Average". A moving average calculates the averages of different subsets within the full dataset, smoothing out the original data in the series. The moving average is commonly used for analyzing data spread over time to "smooth" momentary fluctuations.

IV.   The plotted trendline depends on the selected range. The range determines how many data points contribute to the average each time. In this case, the X-axis is measured in years, so each unit is one year. Enter '8' in order calculate the average changes over two consecutive full government tenures.

Questions

11.   Examine the trendlines in Figure 5.5: Do you identify any historical patterns?

The two trendlines reflect opposite trends. While the average size of an Israeli coalition slightly decreased over time, the average tenure of the government increased by almost 10 months.

Let us return to the question with which we opened this case study: is there a correlation between the parliamentary support of a coalition and its tenure? It is difficult to say that there is consistency between the two metrics. In fact, Yitzhak Galnoor and Dana Blander claim that, "if we examine the coalitions formed throughout the history of Israeli parliamentary politics ... [it seems that only] a model of ideological consistency would predict the type of coalition that emerges" (Galnoor and Blander 2013, 597). Although we did not solve our initial question, we have raised a new and interesting assumption regarding the political history of the State of Israel. Over time, Israeli governments tend to be based on smaller coalitions and serve longer tenures. We will not provide any explanation for this result; we leave the search for an explanation to you.

5.2.2 Coalition stability II

"Modern democracy is unthinkable save in terms of political parties", famously claimed Elmer Schattschneider (1942, I). Indeed, during the pre-statehood period and the early years of the State of Israel, party politics was the essence of the political system.

In contrast to the Israeli multi-party system, there are democracies like Britain, the United States, and Ghana where two main parties dominate, regularly winning an absolute majority of seats in the legislative authority. In 1890, Rand McNally, a publisher specializing in various cartographic publications, released a political history poster designed by Houghton. This poster provides many fascinating details about the political history of the United States, as of 1890, but the most impressive part is the river-like diagram appearing below a timeline at the top of the map (Figure 5.6).

The diagram depicts the American party system from the founding of the United States in 1776 until 1890. In this chart, each ribbon winding upward represents a particular party, and its height indicates the relative share of that party in the American Congress. The diagram illustrates how the Federalist Party transformed into the Democratic-Republican Party around 1820 and how the Republican Party split into two new parties around 1828. The message conveyed by this diagram is that the two-party reality in the United States is dynamic, constantly changing and evolving.

Similarly, we would like to map the Israeli party system, but the continuous changes in the number of parties and their compositions would make such a mapping much more complex than the Houghton diagram. In the words of Benjamin Neuberger (2001, 79): "The party system is so complicated that 'you can't see the forest for the trees.'" Therefore, we will only focus on the three most dominant parties in each election, disregarding their names or compositions.

Questions

12. Our data includes the number of representatives won by each of the three prominent parties in Israel's 25 election cycles since the state's establishment (data available at https://zefsegal.com/computational-liter acy-for-the-humanities/). Which type of chart would best illustrate the relative size of the prominent parties across election cycles? Experiment with different types.

As we have already observed, comparing between 25 different pie charts is a challenge. A scatter plot would be equally challenging since the resultant diagram

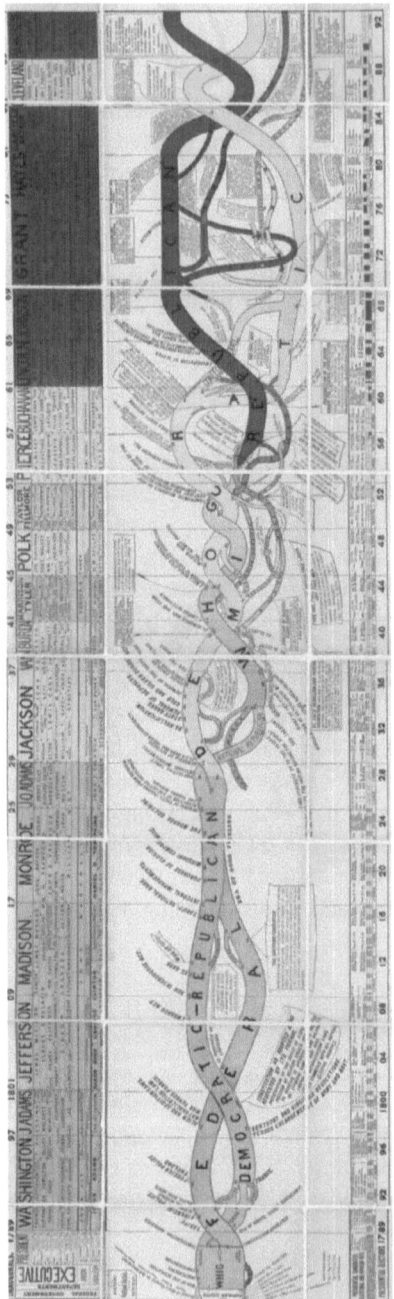

Figure 5.6 A political timeline of American party dominance designed by Walter Houghton (1890).

would include 75 different points. A line chart would enable an exploration of each level separately – major, minor, and tertiary parties – but wouldn't provide a clear comparison between cycles. A stacked bar chart will probably be the best choice since it enables a comparison between each cycle as well as between the different levels of political dominance.

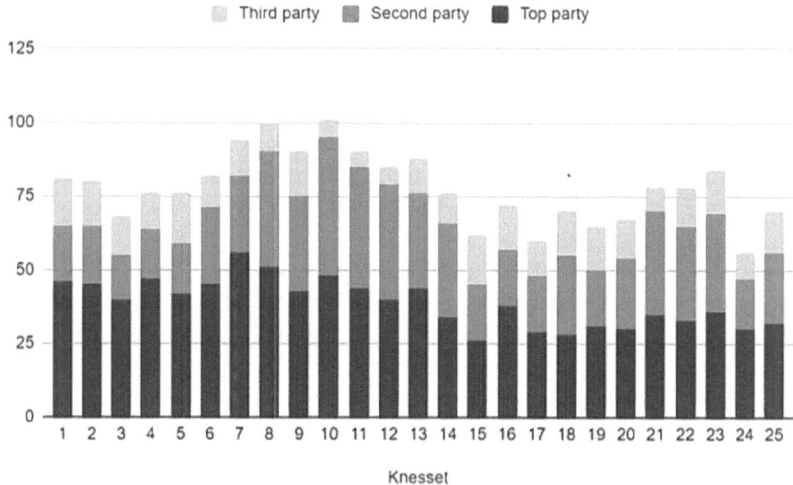

Figure 5.7 A stacked bar chart of the three most dominant parties in each election cycle in Israel (1948–2023).

Questions

13. *Based on the diagram in Figure 5.7, in which election cycles do you identify turning points?

Figure 5.7 reveals a number of significant political events in the history of the state of Israel. The elections for the third Knesset in 1955 were conducted amidst several scandals that cast a shadow on Mapai, the ruling party at the time: the incumbent prime minister Moshe Sharett was dismissed and replaced by David Ben-Gurion; the party was accused in a legal trial of ignoring the plight of European Jewry during the Holocaust; and there was an ongoing conflict between the party and its main political ally, the General Zionists. As a result of these events, the power of the two main parties decreased, while the major opposition party, Herut, doubled its strength.

The elections for the eighth (1973) and tenth (1981) Knessets were similar due to public criticism toward the ruling party. The 1973 elections took place shortly after the 1973 War and the subsequent public discontent. Although there was a slight decrease in the power of the first and third party (Hama'arakh and the National Religious Front), the Likud party significantly increased its strength. The 1981 elections did not follow a war but occurred after the signing of the peace agreement with Egypt. Rising inflation and corruption scandals led to a turbulent election, in which both the Likud and Hama'arakh gained almost 50 seats each. Overall, the period between the eighth Knesset (1973) and the end of the thirteenth Knesset (1996) was characterized by a two-party struggle between both parties.

In 1992, Israel adopted a system of direct election of the prime minister in an attempt to produce more stable governments. The 1996 election results (Knesset 15) reflected this change, as the power of the major parties decreased, and the smaller parties became stronger. Even though the previous electoral system was reinstated in 2001, the change in the voting pattern of Israeli voters continued in subsequent elections.

The results of the 21st, 22nd, and 23rd Knesset elections, and the rise in the power of the major parties, resemble to some extent the election systems of the mid-1980s, especially the inability of each of the major parties to form a coalition independently without mutual cooperation. The two major parties, Likud and Kakhol-Lavan, positioned themselves as representatives of each bloc, attracting votes that were previously given to smaller parties.

In conclusion, the diagram we created provided us with an overview of the Israeli electoral system, allowing us to analyze patterns of change in Israeli party politics. Identifying pivotal points in the chart and correlating them with contemporary political events enabled us to conduct a deeper analysis of the political system. It's important to emphasize that the chart is not an explanation of historical changes but rather a tool that allows us to focus on specific events. The explanation itself requires explicit reference to the events themselves.

5.3 Case study: Word frequencies in literary studies

We have already experimented with the initial steps of text analysis in Chapter 3. We learned how to (1) "clean" a text from punctuation marks and convert all the English letters to lowercase; (2) divide a text into words and calculate the number of tokens, types, sentences, and quotation marks; and (3) use these findings to compare between two short stories. We received the following results:

Story	Tokens	Types	Sentences	Quoted texts
The Haunted Mind	1,791	787	52	1
Two Kinds	4,547	1,328	319	70

"The Haunted Mind" by Nathaniel Hawthorne and "Two Kinds" by Amy Tan exhibit very different profiles. "Two Kinds" is longer than "The Haunted Mind" and, appropriately, has a larger vocabulary (1,328 types compared to 787). Nevertheless, vocabulary richness is not only a function of the size of the vocabulary, but also of the rate of repetitions. The relatively higher type–token ratio in "The Haunted Mind" (0.44), compared to "Two Kinds" (0.29), indicates a richer vocabulary.

The next step in text analysis focuses on the words themselves, starting with their frequency in the text. To do this, we will use Python to create a list of types in each text, examine the distribution of frequencies, and generate a line diagram illustrating it. Before doing so, let's familiarize ourselves with a data structure called "dictionary".

5.3.1 Dictionaries in Python

A dictionary in Python is a data structure that stores its items as ordered pairs separated by a colon, in which the first is called a "key" and the second a "value". An example of a dictionary is,

```
works = {"Mozzini": 35, "Dubyk": 15, "Eskandari": 5,
"Buffik": 2}
```

The formal distinction between a list and a dictionary is primarily expressed in the type of brackets used: a list appears between square brackets, while a dictionary appears between curly brackets. Similar to a list, the elements of the dictionary appear one after the other, separated by commas. However, unlike a list, there is no significance to the internal order of the elements within the dictionary. Therefore, accessing values is not done by positional index but rather by using the key. We'll practice accessing dictionary items in the next programming exercise.

The works dictionary described above contains a set of female artists, and for each artist, the number of their works is specified (according to WikiArt). Here, the key is the painter's surname (a string), while the value is the number of their works (a numeric value). In works, each dictionary item maintains the same relationship, and in this case, the value indicates the number of artworks for the corresponding key. This is just one example, and in practice, you can represent various types of information in a dictionary. Thus, the key can also be a number, and the value can be of any type (integers, lists, booleans, tuples, etc.). See, for example, the following dictionary:

```
Artist = {"Surname": "Mozzini", "First Name": "Jessica
Alves   Oliveira",   "Born":   1999,   "Works":   35,
"Style": "Post-Impressionism", "In WikiArt": True}
```

Table 5.1 A dictionary represented in table format

Surname	Mozzini
First Name	Jessica Alves Oliveira
Born	1999
Works	35
Style	Post-Impressionism
In WikiArt	TRUE

The "artist" dictionary contains various pieces of information about the painter Jessica Alves Oliveira Mozzini. Each item in the dictionary stores data of a different type, described by the key (name, birth year, etc.). Some values in "artist" are strings, some are integers, and the last value is boolean.

The type of information that is stored in a dictionary can be represented in a table format:

As will be evident later on, a dictionary is an exceptionally useful data structure for processing information in Python.

Creating a dictionary in Python

In the following code block, the dictionary abbrev is defined, which contains several English acronym expansions. Each item in the dictionary consists of a key, in this case, the acronym, and a value – the expansion. The two types of data here are represented using strings.

```
abbrev = {"ROFL": "Rolling on floor laughing",
"ICYMI": "In case you missed it", "TL;DR": "Too long,
didn\'t read", "LMK": "Let me know", "NVM": "Nevermind",
"TBH": "To be honest"}
```

Please note the word 'didn't' in the third entry in the dictionary. The word contains an apostrophe, which is a reserved character in Python strings. To ensure that Python treats it as a regular character and not as a reserved one, a backslash `\` is added before it.

To add an additional item to the dictionary, one needs to specify the key and assign the corresponding value to it.

```
abbrev["BTW"] = "By the way"
```

Retrieving dictionary items

There are a number of Python methods that allow you to retrieve the contents of a dictionary:

```
items()     Returns a list of all key–value pairs
keys()      Returns a list of the keys
values()    Returns a list of the values
```

Since there is no inherent order to the items in a dictionary, we cannot use an index to retrieve an item, as we did in the case of a string (e.g., `mixed_list[0]`). Instead, accessing an item is done using the key.

```
abbrev["NVM"]
```

Before executing the next code, what do you think will be printed on the screen?

```
ab = input ("Type an abbreviation.")
define = abbrev[ab]
print (ab + " means " + define)
```

As we've seen Chapter 3, the `input()` function allows us to receive user input and perform operations on it. In this case, we use the input as a key to retrieve a value from the dictionary and then print a message.

```
Type an abbreviation.TBH
TBH means To be honest
```

However, if our user's input is not present in the dictionary, our program "crashes" and we receive a "KeyError" message, indicating that there is no such key in the dictionary.

```
KeyError: 'LOL'
```

When writing software, the aspiration is to address all possible scenarios, thus avoiding "crashes" (i.e., program interruptions). In the case of retrieving values from a dictionary, Python provides the `get()` method, which takes the key as an argument and returns the corresponding value if it exists. When the key is present, direct retrieval (with the key in square brackets) and retrieval using `get()` are equivalent.

However, when the key is not present, the `get()` method returns the value `None` and the program does not crash.

```
print(abbrev.get("LOL"))
```

```
None
```

The value `None` (not the string "None") is a special value in Python that is returned by a function when it does not return any other value. This value is very useful for preventing software "crashes" in cases of failure and for handling errors appropriately. Consider, for example, a situation where you enter an incorrect username or password in an application.

Questions

14. *Complete the following code using the value returned from `get()`. If it is equal to `None`, print an error message; otherwise, print the returned value.

```
ab = input ("Type an abbreviation.")
define = abbrev.get(ab)
if
else:
```

Iterating over dictionary items

We learned how to retrieve and print all the items in a dictionary. However, typically, we don't want to print the entire dictionary, but rather go through each item and perform some action. We can do this using a `for` loop. As we've seen in Chapter 3, a `for loop` can iterate over all the items in iterable data objects.

With strings, the loop iterates character by character.

```
# print character by character
for i in "hello world":
    print(i)
```

With lists, the loop iterates character by character.

```
mixed_list = ["red", 12, "John Lennon", "Ukraine",
58,"y"]
# print list item by list item
for i in mixed_list:
    print(i)
```

Iterating over items in a dictionary is similar to iterating over strings and lists but slightly more complex. What do you expect will be printed on the screen?

```
for i in abbrev:
    print(i)
```

In each iteration, the variable i stores a different key from within the dictionary.

Questions

15. *Try modifying the code so that it prints the values in the dictionary one after the other, instead of the keys.

Consider: Why is there no need to use the get() method in this case?

To iterate over a dictionary and retrieve both the keys and the values, we will use the items() method, which we used previously to retrieve all the key–value pairs and iterate with two different variables. In each iteration, we assign the key to the variable k and the value to the variable v.

```
for k, v in abbrev.items():
    print(k + " means " + v)
```

5.3.2 *Creating word frequency lists*

In this section, we will use a dictionary variable to create a word frequency list for the words in a text. However, before we create the list we must first go through the text preparation steps that we learned in Chapter 3, namely text normalization and splitting the text into a list of tokens. In the following example, we see the raw text stored as string and, following it, the list of tokens that is created after the raw text has been "cleaned" of punctuation marks and all uppercase letter have been converted to lowercase.

```
" 'What day is it?', asked Winnie the Pooh. 'It
is today,'squeaked Piglet. 'My favorite day,'
said Pooh."
['what', 'day', 'is', 'it', 'asked', 'winnie', 'the',
'pooh', 'it', 'is', 'today', 'squeaked', 'piglet',
'my', 'favorite', 'day', 'said', 'pooh']
```

The frequency list which corresponds to this text is given in the dictionary below.

```
{'what':   1,   'day':   2,   'is':   2,   'it':   2,
'asked': 1, 'winnie': 1, 'the': 1, 'pooh': 2, 'today':
1, 'squeaked': 1, 'piglet': 1, 'my': 1, 'favorite': 1,
'said': 1}
```

The dictionary consists of pairs: a word (the key) and its frequency in the text (the value). Thus, for example, in our text, the word 'what' appears once, and the word 'day' appears twice. The number of items in the dictionary is equal to the number of types, i.e., the number of different words in the text. The sum of all the values in the dictionary is equal to the number of tokens in the text, i.e., the total number of words.

The algorithm for creating a frequency list is as follows:

Normalize the text and create a list of tokens.
Initialize a variable `freqs` as a empty dictionary.
Iterate over the list of words in the variable `tokens` word by word.
If the word is already in the dictionary:
    Add 1 to its value in the dictionary.
Otherwise:
    Add a new entry to the dictionary: the key is the word, and the value is 1.

Questions

16. Creating a frequency list
    a. Prepare the two short stories explored in section 3.9, "The Haunted Mind" by Hawthorne and "Two Kinds" by Amy Tan, by normalizing them and creating tokens lists.
    b. *"Translate" the previously ascribed pseudo-algorithm into Python code and run the program.

In order to make the resultant dictionary more informative, we want to sort it by descending order of frequency. However, unlike a list, the items in a dictionary do not have any specific order. We can think of this data structure as a bag full of items. Therefore, to sort the items, we need to convert the dictionary into a list of tuples (a word and its frequency) and then apply the `sort()` method to it. However, when the `sort()` method is applied to a list of tuples, it sorts the list based on the value of the first item in the tuple. Since we want to sort the list based on the frequency of the words, we will copy the dictionary items to the list in reverse order. First the value (frequency) and then the key (word).

```
# dictionary
{'what':    1,    'day':    2,    'is':    2,    'it':    2,
 'asked': 1, 'winnie': 1, 'the': 1, 'pooh': 2, 'today': 1,
 'squeaked': 1, 'piglet': 1, 'my': 1, 'favorite': 1,
 'said': 1}

# list of tuples (frequency, word)
[(1, 'what'), (2, 'day'), (2, 'is'), (2, 'it'),
 (1, 'asked'), (1, 'winnie'), (1, 'the'), (2, 'pooh'),
 (1, 'today'), (1, 'squeaked'), (1, 'piglet'), (1,
 'my'), (1, 'favorite'), (1, 'said')]
```

As you can see in the example above, dictionaries and lists are distinguished by the shape of their brackets – curly for dictionaries and square for lists. In both cases, items are separated by commas, but the items of the dictionary are presented as "key: value" pairs, while the items of the list are tuples enclosed in parentheses and separated by commas.

The reversed copying process is implemented in the following code.

```
freqs_l = []
for k, v in freqs.items():
    freqs_l.append((v, k))
```

First, it assigns an empty list value to the variable `freqs_l`. This is a necessary step, since it is impossible to append items into a list variable that has not been defined yet. Next, it iterates over the output of the `items()` method, applied to the dictionary `freqs`. As we saw previously, `items()` returns the dictionary items as pairs – key and value. Therefore, in the `for` loop, we use two variables, `k` (for the key) and `v` (for the value). In each iteration, we add a tuple `(v, k)` to the freqs_l list using the `append()` method. We reverse the order of the elements in the tuple, relative to their order in the dictionary.

To sort the list we call the `sort()` method. Note that this method does not return a new list but sorts the original one. In order to sort the list in descending order of frequency we add the argument `reverse=True`.

```
freqs_l.sort(reverse=True)
```

```
[(2, 'pooh'), (2, 'it'), (2, 'is'), (2, 'day'), (1,
'winnie'), (1, 'what'), (1, 'today'), (1, 'the'),
(1, 'squeaked'), (1, 'said'), (1, 'piglet'), (1,
'my'), (1, 'favorite'), (1, 'asked')]
```

Questions

17. Frequency lists
    a. Sort the frequency lists of the two stories.
    b. *Which are the three most frequent words in each story?

5.3.3 Analyzing frequency lists

In the previous section, we created word frequency lists, sorted them, and then printed them on the screen. However, it would be much more convenient to analyze the lists in a spreadsheet. To do so, we will first save them as files in a CSV (comma-separated values) format, in which the values for each data point appear on a separate line, separated by commas.

In the following code block, we iterate through `freqs_l`, the list of tuples which we previously created, and write each tuple in a CSV format into a file.

```
output_file = open("frequencies.csv", "w")
for freq, item in freqs_l:
    output_file.write(str(freq) + ", " + item + "\n")
output_file.close()
```

```
2, pooh
2, it
2, is
2, day
1, winnie
. . .
```

Notice the following elements in this code:

- The w argument of open() indicates that the file is opened for writing.
- Inside the for loop, for each tuple in the list, the write() method is used to write the tuple into the file.
- When writing the tuple the function str() is used to convert the integer freq to a string.
- At the end of the tuple a line break "\n" is added.

Note that a CSV file is not a spreadsheet and cannot be simply opened in Google Sheets. Instead, in Google Sheets, select "Import" in the "File" menu. At this point you can either select a CSV file from your Google drive or choose "Upload" and select the CSV file from your local computer. When prompted, make sure to configure the import settings so that the separator is defined as a comma.

Questions

18. Create output files with the frequency lists for both stories.
19. Import both csv files into Google Sheets. When importing the second file, mark "Insert new sheet(s)" and not "Create new spreadsheets".
20. Analyze the results
    a. *Which of the top ten words are common to both texts and which of the top ten words are unique to each text?
    b. *What do the ten most frequently used words tell us about the content of the stories?
    c. *Examine the complete frequency lists. What would you guess is the content of each story? Which words did you use to make your assumptions?

Before we proceed to create a chart, let's examine the frequency lists we generated. These lists of words and their frequencies provide us with an "overview" of the texts' content.

Frequent words in a text can be divided into two types: words that are generally common and words that are specific to a given text. The first type includes function words such as 'the', 'and', 'of', and 'a', commonly known as **stop words**. These words do not reveal the content of the text, and are therefore often ignored. However, sometimes these words help distinguish between different writers, regardless of the content of their writing. Words that are specific to a given text assist in identifying the content of the text. Even non-frequent words that appear toward the end of the frequency list can serve as clues about the text's content.

5.3.4 Zipf's law

In Chapter 3, we calculated the type–token ratio for each story. This ratio reflects the text's lexical richness, enabling us to distinguish between different texts. However, there is considerable variation in the frequency of individual words in the text and the ratio does not necessarily capture the entire story. An additional perspective on texts and their vocabulary is provided by the distribution of frequencies in the text.

The data we generated using Python is now presented in two columns of the spreadsheet, sorted in decreasing order by frequency. Using this data, we can create a chart that plots the rank of a word (or type) in the frequency list. In other words, the X-axis will indicate the rank, and the Y-axis will represent the frequency. The word itself, appearing in the second column of the spreadsheet, will not be depicted in Figure 5.8.

Surprisingly, despite the many differences between both stories, the diagrams are very similar. In fact, an analysis of most texts would result in similar curves. This is a statistical phenomenon called "Zipf's law", named after the linguist George Kingsley Zipf (1902–1950), a pioneer in computational linguistics. Zipf's law describes the relationship between words in a text and their frequency. According to the law, the frequency of the most common word in a large text is approximately twice that of the second word in the list, three times that of the third word, and so on. As can be seen in the diagrams, the curve describing the relationship initially drops steeply and quickly "flattens" into a horizontal line close to the X-axis, known as the "long tail". Zipf's law is universal and holds for most natural languages, and text of any type and length, from short poems to encyclopedias.

The implication of Zipf's law is that in every text, there are very few highly frequent words, a larger number of moderately frequent words, and many words that appear only once (also known as **hapax legomena**). In long texts, approximately 40% and 60% of the words are hapax legomena.

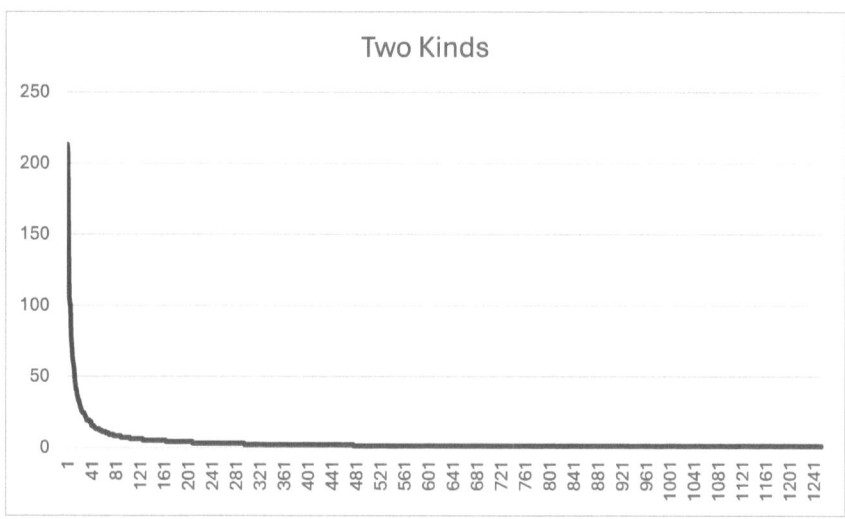

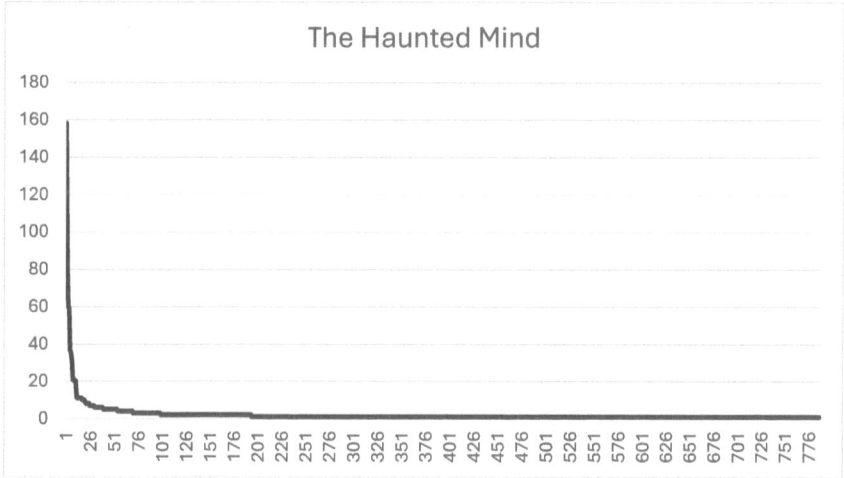

Figure 5.8 Line charts depicting the rank to frequency distribution of "Two Kinds" and "The Haunted Mind".

Questions

21. Create two line charts illustrating the frequency relative to the position in Amy Tan's and in Nathaniel Hawthorne's stories.
22. Analyze the data

a. *Examine the charts and roughly estimate, for each story, the percentage of words belonging to the "long tail" of the curve.
b. *Examine the spreadsheets and calculate the percentage of hapax legomena in each story.

The higher percentage of hapax legomena in "The Haunted Mind" (75% of all types compared to 62% in "Two Kinds") corresponds to the relatively high type–token ratio in Hawthorne's story and reflects its lexical richness. This is also true when looking at the "long tail" of both stories, as 91% of words in "The Haunted Mind" appear less than four times compared to 83% of the words in "Two Kinds". The remaining words have more variation and most likely more significance, but due to display constraints, their part of the diagram appears "compressed", making it difficult to discern the data.

5.3.5 Distant and close readings of literary texts

By creating frequency lists, we compared two stories, incorporating two perspectives: a distant view and a close view. The distant view allowed us to examine the distribution of word frequencies in both texts using a curve. Interestingly, despite the differences between the texts, the distribution of frequencies is relatively similar. As mentioned previously, Zipf's Law, describing the distribution, is a general law that applies to any text, regardless of its content or length.

Examining texts from a distant perspective is a method developed by the Italian-American researcher Franco Moretti (2013), known as "distant reading". In distant reading, computational and statistical tools are used to create a simplified picture of one text or, more often, a selection of texts, without actually "reading" them. Moretti describes it as "a little pact with the devil: we know how to read texts, now let's learn how not to read them".

Figure 5.9 illustrates the results of a distant reading conducted on a broad corpus of 1,117 books published between 1800–1900.

The X-axis represents time, oscillating between 1795 and 1900. The Y-axis reflects the **median type–token ratio** (median TTR). The median ratio, as opposed to the simple ratio that we calculated in Chapter 3, is a way to overcome the text size effect, namely that longer texts tend to have lower TTRs. This measure is calculated by first dividing the text into 1,000 token segments and calculating the TTR for each segment. The median TTR is the median of all the values. We will learn about the median in the next chapter.

Each point on the chart represents a book belonging to one of two corpora – the sample representing the canon of 19th-century English literature or a sample from the archive of the University of Illinois. Each point is positioned at the intersection of the publication year of the book and the median TTR. As seen in the chart, there is a decline in the ratio over the years, indicating a decrease in linguistic richness.

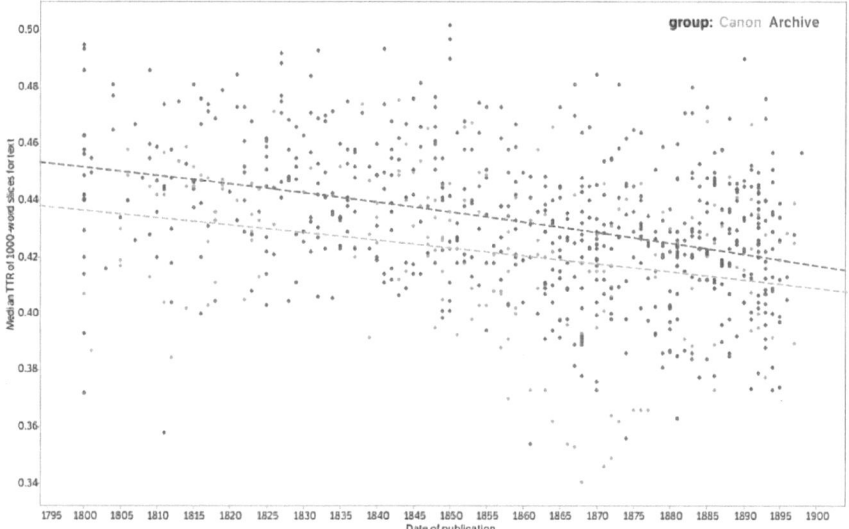

Figure 5.9 The median TTR of 1,117 books over time.

Source: Algee-Hewitt et al. (2016).

The "close" analysis we conducted is not a classic literary analysis. We didn't actually read the text; instead we used computational tools to represent it differently. When examining the lists of frequent words in each story, we observed that, despite being "just" lists, they do reflect a certain aspect of the text. The frequent words in a specific text (those that are not merely stop words) serve as clues about the content and writing style (personal, lyrical, informative, etc.) of the text.

It is worth noting that the idea of distant reading has not been universally embraced in literary research. Supporters argue that the ability to view texts in a different way exposes aspects not noticeable to close-reading scholars. These aspects can serve as a starting point for innovative literary research. Opponents view it as an unwarranted and inappropriate intrusion of exact sciences into a field that is not theirs, producing research that addresses questions of little value to literary researchers.

Glossary

Big Data Large datasets that are too vast to be processed or analyzed with traditional methods. In the humanities, big data involves dealing with large-scale collections of texts, images, or other cultural artifacts.

Data Mining The process of extracting useful information from large datasets by identifying patterns, correlations, or anomalies, often with the help of computational tools.

Dictionary A Python data-structure that stores data in key–value pairs, allowing fast access to values using unique keys.

Hapax legomena Words that appear only once in a text.

Small Data Datasets that are manageable in size and often analyzed manually, as opposed to big data. Small data is typical in many humanities fields, where the focus is on detailed, qualitative analysis.

Trendline A line on a graph that shows the general direction or pattern in a set of data points over time, used to analyze trends in historical, economic, or social data.

Visualization The process of representing data or information visually through charts, graphs, or diagrams, enabling clearer analysis and communication of complex datasets.

Zipf's Law A principle that states that in any large sample of natural language, the frequency of any word is inversely proportional to its rank in the frequency table.

Note

1 The numbers do not add up to 517. A few paintings are listed under multiple styles.

References

Addison, Joseph. 1712. *The Spectator*. November 6.

Algee-Hewitt, Mark, Sarah Allison, Marissa Gemma, Ryan Heuser, Franco Moretti, and Hannah Walser. 2016. *Canon/Archive: Large-Scale Dynamics in the Literary Field*. Stanford Literary Lab, Pamphlet 11.

Corot, Jean-Baptiste-Camille. 1850. "The Dance of the Nymphs". Oil on canvas, 98 × 131 cm. Musée d'Orsay, Paris.

Delacour, Hélène, and Bernard Leca. 2017. "The Paradox of Controversial Innovation: Insights from the Rise of Impressionism". *Organization Studies* 38(5): 597–618. https://doi.org/10.1177/0170840616663237

DiMaggio, Paul. 1987. "Classification in Art". *American Sociological Review* 52(4): 440–455. https://doi.org/10.2307/2095290

Galnoor, Itzhak, and Dana Blander. 2013. *The Political System in Israel*, Vol. I. Am Oved.

Gibson-Wood, Carol. 2002. "Picture Consumption in London at the End of the Seventeenth Century". *The Art Bulletin* 84(3): 491–500.

Houghton, Walter. 1890. *Houghton's New Reversible Political and U.S. Map Combined*. Rand, McNally & Co.

Lippincott, Louise. 2013. "Expanding on Portraiture". In *Consumption of Culture*, edited by Ann Bermingham, and John Brewer. Routledge.

Lizardo, Omar. 2008. "The Question of Culture Consumption and Stratification Revisited". *Sociologica* 2: 1–32. https://doi.org/10.2383/27709

McNally, Mark. 2014. "The Marketing Techniques of William Hogarth (1697–1764), Artist and Engraver". PhD Diss., Durham University.

Moretti, Franco. 2013. *Distant Reading*. Verso Books.

Neuberger, Benjamin. 2001. "Between Ideology and Sociology: The Parties in Israel 1950–2000". *Medina ve-Hevra* 1(1): 79–87

Roivainen, Hege, Leo Lahti, Mark J. Hill, Jani Marjanen and Mikko Tolonen. 2019. "Book Formats and Reading Habits in Early Modern Europe". *DataverseNL*. https://doi.org/10.34894/BHX43Q

Schattschneider, Elmer E. 1942. *Party Government*. Holt, Rinehart and Winston.

Whitley, William T. 1928. *Artists and Their Friends in England 1700–1799*, Volume I. Medici Society.

Wittmann, Reinhard. 1999. "Was There a Reading Revolution at the End of the Eighteenth Century?". In *A History of Reading in the West*, edited by Guglielmo Cavallo and Roger Chartier. Polity Press.

Yaakobi, Gad. 1980. *The Government*. Am Oved.

Yanai, Natan. 1987. "The Statist Conception of Ben-Gurion". *Kathedra* 45: 169–189.

Solutions to selected questions

1. Re. Figure 5.1:
   a. It is difficult to say definitively, as the sample is very small with only four artists, but Henri Matisse stands out the most in terms of diversity. Matisse is associated with a much larger array of styles compared to the other artists. Cezanne and Gauguin are very similar, with around 50% of their work identified as Post-Impressionist, and the rest associated evenly with two to three other styles. Henri Martin has a similar number of associated styles but his primary style Divisionism (a characteristic style in late 19th century Neo-Impressionism) takes up a larger portion of his work.
   b. Cézanne, born in 1839, began painting nearly a decade before Gauguin but had already become a friend and mentor to him by the early 1870s. Henri Martin, born in 1860, entered the scene later than both, which accounts for the more distant connection in their artistic styles. Matisse, the youngest of the group, born in 1869, held his first exhibition in 1901. The chronology of their birth years closely mirrors the variety of artistic styles associated with each painter.

4. Despite variations, all regions witnessed a decline in the proportion of folio-sized books (approximately 30 x 48 cm^2) and a corresponding rise in octavo-sized books (approx. 15x 23 cm^2), with duodecimo size (approx. 13 x 19 cm^2) gaining prominence in Britain.

5. Plotting the data:
   a. A scatter chart is a suitable visualization, as it presents each artifact based on two quantitative measures: area and year.
   b. The date and the area columns are the only columns providing relevant data. There is also a possibility to label each artifact by including the first column (Title).

9. A double-bar chart would display columns of different metrics side by side, making it challenging to observe changes over time.

13. Unlike the chart we created in the previous section, the current one shows clear patterns of change. A simple way to identify turning points is looking for extrema (peaks and troughs). We find trough points in the 3rd, 15th, and 17th cycles, while the peak points are the 8th, 10th, and 23rd election cycles.

14. The complete code

```
ab = input("Type an abbreviation.")
define = abbrev.get(ab)
if define == None:
    print(ab + " is not an abbreviation in the
    dictionary.")
else:
    print(ab + " means " + define)
```

15. The modified code

```
for i in abbrev:
    print(abbrev[i])
```

16. b. From a token list to a frequency list

```
freqs = {}
for word in tokens:
    if word in freqs:
        freqs[word] = freqs[word] + 1
    else:
        freqs[word] = 1
```

17. b. The three most frequently used words in "The Haunted Mind" were 'the', 'of', and 'a'. while the three most frequently used words in "Two Kinds" were 'I', 'the', and 'and'.

20. a. The following top ten words are common to both stories: 'the', 'of', 'a', 'and', 'to'. The following top ten words are unique to "The Haunted Mind": you, in, your, with, that. The following top ten words are unique to "Two Kinds": 'I', 'my', 'was', 'she', 'me'.

   b. "Two Kinds" is narrated in the first person ('I', 'my', 'me') and refers to another female character. "The Haunted Mind" is narrated in the second person (you, your).

   c. It seems that "Two Kinds" describes a relationship between a mother and daughter, related to music and piano lessons. The story likely addresses the question of musical talent (whether the daughter has it or not), and it may depict an event of playing in front of an audience. This assumption is based on the following relatively frequent words: 'I', 'mother', 'piano', 'play', 'tv', 'music', 'Chinese', 'prodigy', 'father', 'notes', 'talent', 'genius', 'lessons', 'parents', 'mothers', 'piece', 'child', 'boy', 'audience'.

The story "The Haunted Mind" appears to describe, in the second person, an experience related to sleep. It might involve dreams that interfere with sleep or perhaps sleep disturbances and situations between wakefulness and sleep. This assumption is based on the following relatively frequent words: you, mind, hour, heart, eye, dream, sleep, moment, sound, slumber, obscurity, life, night, heavy, head, darkness, clock, awake.

22. a. In the story "Two Kinds", the curve levels off around word 211. The number of tokens is 1,254, so the percentage is approximately 80%.
   In the story "The Haunted Mind", the curve levels off around word 109. The number of tokens is 787, so the percentage is approximately 86%.

   b. In the story "Two Kinds", there are 778 hapaxes, which constitute 62% of the tokens. In the story "The Haunted Mind", there are 778 hapaxes, which make up 75% of the tokens.

6   Introduction to descriptive statistics

This chapter begins with practical examples demonstrating the use of statistics in historical and literary research, setting the stage for understanding core statistical concepts. Each example ends with questions, which will be addressed in the following chapters.

Between 1881 and 1914, approximately 1.5 million Jews emigrated from the Russian Empire to the United States, a migration often attributed to waves of pogroms in southern Russia. However, statistical analysis reveals a more nuanced picture. Research based on over 5.7 million immigration documents demonstrated that there was no dependence between the demographic and geographic composition of the immigrants and the location of these disturbances (Spitzer 2021). How was this dependency measured?

In 1851, the British mathematician Augustus De Morgan (1806–1871) claimed that the authenticity of St. Paul's writings could be established by measuring the length of the words used in his Epistles (de Morgan 1851). In 1887, the American physicist Thomas Mendenhall (1841–1924) tested De Morgan's hypothesis to address the accusation that Francis Bacon (1561–1626), the English philosopher, was the author of the plays attributed to Shakespeare (Mendenhall 1887). To perform the test, Mendenhall measured the length of 200,000 words from Bacon's writings and 400,000 words from Shakespeare's writings. Although his research showed that word length was not a reliable measure for identifying an author, the discovery of remarkably similar stylistic features between Shakespeare's writings and those of his contemporary, Christopher Marlowe (1564–1593), continues to fuel literary research to this day. But what is a "reliable measure for identifying an author" and what can be considered "similar stylistic features"?

Life expectancy is a measure used to describe different societies in both the past and present. Table 6.1, for example, lists the life expectancy estimates for various parts of the world in 1850, 1900, and 1950. At a glance, it is evident that during this period, life expectancy increased dramatically. In Asia, life expectancy rose from 27.5 years in 1850 to 41.6 years in 1950, and in Europe, it increased from 36.3 years to 64.7 years, respectively. Does this mean that in the mid-19th century, there were no elderly people in Europe and Asia? Of course not. Jan Tzatzoe, a Xhosa chief in South Africa, died in 1868 at the age of 76. Queen Victoria of

DOI: 10.4324/9781003502814-7

Table 6.1 Life expectancy in different continents between 1850 and 1950

Year	Oceania	Europe	America	Asia	Africa
1850	34.7	36.3	35.1	27.5	-
1900	47.6	42.7	41	28	26.4
1950	63.4	64.7	58.4	41.6	35.6

Source: Riley (2005).

England died in 1901 at the age of 82. Rabbi Yahya Kafah, one of the greatest sages of Yemen, passed away in 1931 at the age of 81. Can we learn anything from these examples? And how do they fit with the life expectancy presented in Table 6.1?

These and other similar studies use samples, population censuses, and surveys to generate quantitative data, which is then further analyzed. But can this data, or more precisely, the results of its analysis, be trusted?

On November 9, 1965, a power outage occurred in the New York City area, leaving the homes of about 30 million people in darkness for many hours. Nine months later, a headline in *The New York Times* read, "Births Up Nine Months After the Blackout". The article described the conclusion-drawing process:

> A sharp increase in births has been reported by several large hospitals here, 9 months after the 1965 blackout. Mount Sinai Hospital, which averages 11 births daily, had 18 births on Monday. This was a record for the hospital; its previous one-day high was 18. At Bellevue there were 29 new babies in the nursery yesterday, compared with 11 a week ago and an average of 20.
>
> (Tolchin 1965)

The article's author, Martin Tolchin, clarified that similar data were present in all hospitals in the area. However, five years later, the American sociologist J. Richard Udry published a three-page article in the journal *Demography*, categorically refuting the claim made by *The New York Times*. Udry argued: "A comparison of the number of births in New York City nine months after the Great Blackout of 1965 with comparable periods for the previous five years shows no increase in births associated with the blackout" (Udry 1970).

Udry based his conclusion on government data on pregnancies and births, stating that 90% of all pregnancies that began on November 10, 1965, ended in births between June 27 and August 14, 1966. Therefore, the 1965 blackout should have affected the entire mentioned period, not just a specific day when a birth increase was observed. However, examining the data showed no significant difference between the number of babies born during this period in 1966 and the number of babies born in each of the five years preceding it. The mathematical reasoning was sound, but Udry was not naive and added: "let us not imagine that a simple statistical analysis such as this will lay to rest the myth of blackout babies". The numbers in the article "competed" with the numbers presented by *The New York*

Times, and it seemed to the general public, who did not delve into the mathematical explanation, that both arguments were equivalent.

A famous quote from the late 19th century claims that "there are three kinds of lies: lies, damned lies, and statistics". This saying reflects the skepticism about the explanatory power often attributed to statistical data. While it is true that statistics can sometimes be misleading, statistical analysis remains crucial for understanding complex phenomena. As we have discussed in previous chapters, a prudent and critical approach to quantitative data – one that includes comprehending the meaning of mathematical operations and recognizing potential biases – can mitigate the risk of misinformation and provide valuable insights into complex social behaviors and trends.

6.1   What is statistics?

Today, statistics encompasses two primary meanings:

- A series of numbers systematically collected by social and political institutions to characterize and analyze natural or social phenomena.
- A mathematical branch dealing with the collection, processing, analysis, and presentation of quantitative data. It provides experimental sciences with uniform quantitative tools and explains the logic and methodology that allow conclusions to be drawn from experiments, despite the high uncertainty involved.

The term "statistics" (*Statistik*) first appeared in the writings of the German economist and jurist Gotfried Achenwall (1719–1772). Achenwall used it to describe verbal, non-quantitative information documenting the social, economic, and political characteristics of different populations. The word originated from the Italian *statista* (statesman), because, from Achenwall's and other statisticians' perspectives of the time, this information was essential for state control and administration.

The quantitative roots of statistics lie in probability studies conducted by Pierre de Fermat (1607–1665) and Blaise Pascal (1623–1662) in the 1650s. Starting in the late 17th century, probability researchers who had previously focused on games of chance began to turn their attention to probabilistic models of uncertainty in society and science. The use of these methods spread from astronomy and geodesy to psychology, biology, and social sciences. While the interpretation and contextualization of statistical analysis depend on the subject matter, its mathematical reasoning remains consistent.

From the 19th century onwards, the two original meanings of statistics – institutional data collection and the science of quantification – merged within European states. Led by France, Britain, and the German states, these countries established statistical offices to understand and analyze the nature of society through numerical values. Statistics transformed from a descriptive tool that provided verbal information about a state, province, or region into a method of data collection and analysis using summation and averages. Its use quickly expanded to civil societies and academics seeking to understand social phenomena through data collected in government population censuses or independent frameworks.

Due to the practical and historical links between statistical analysis and data collection, statistics are never neutral. Every data collection includes biases that reflect the values of those who collected the information. For example, the British population census has been conducted every decade since 1801 (except for 1941), but the questions have changed to reflect societal changes and government needs (Canning 2014, 8). Between 1951 and 1991, respondents were asked if their homes had toilets. This question, quite important in 1951 due to the lack of infrastructure in many parts of Britain, became redundant by 1991. Since 2011, respondents have been asked about their level of English, a question that had never appeared in previous censuses despite Britain absorbing significant waves of immigration before 2011.

In the US census, racial categories reflect the changing racial biases of Caucasian American society, as they have always been predetermined by the central statistics bureau and altered almost every decade (Pew Research Center 2020). In 1890, categories like "quadroon" (one-quarter black and three-quarters white) and "octoroon" (one-eighth black and seven-eighths white) were added to distinguish between different levels of racial mixing. However, in 1930, these categories were abolished because a new guideline stated that any descendant of white and black mixed heritage would be considered black "no matter how small the percentage" of black blood. Native Americans did not appear as a separate category until 1860, and in 1890, those living on reservations were distinguished from those living outside reservations. In 1960, the census introduced categories for Eskimos and Aleuts, but only in Alaska. Since 2000, all indigenous categories have been combined under "American Indian or Alaska Native," with an option to note the respondent's tribal affiliation.

Additionally, in 1960, respondents were first allowed to choose their own race, and since 2000, they have been allowed to select more than one racial category. These changes highlight how census categories are not neutral but instead mirror the prevailing racial biases and discriminatory practices of the hegemonic society.

These examples indicate that databases should always be critically examined. It is necessary to ask: What was the purpose of data collection? What is the logic behind the different categories?

6.2 Central tendency measures

One of the main goals of modern statistics is to provide a concise and easily comprehensible description of numerical data. In 1835, the Belgian statistician Adolphe Quetelet (1796–1874) coined the term "average man" (*l'homme moyen*), whose traits are the average of the physical, economic, and social traits in each country (see Figure 6.1). According to Quetelet, this person represented the entire population of the country and the characteristics of society. This perception is undoubtedly problematic as it links mathematically average attributes to an idealized personification of society, thus blurring the existing diversity within a society. Nevertheless,

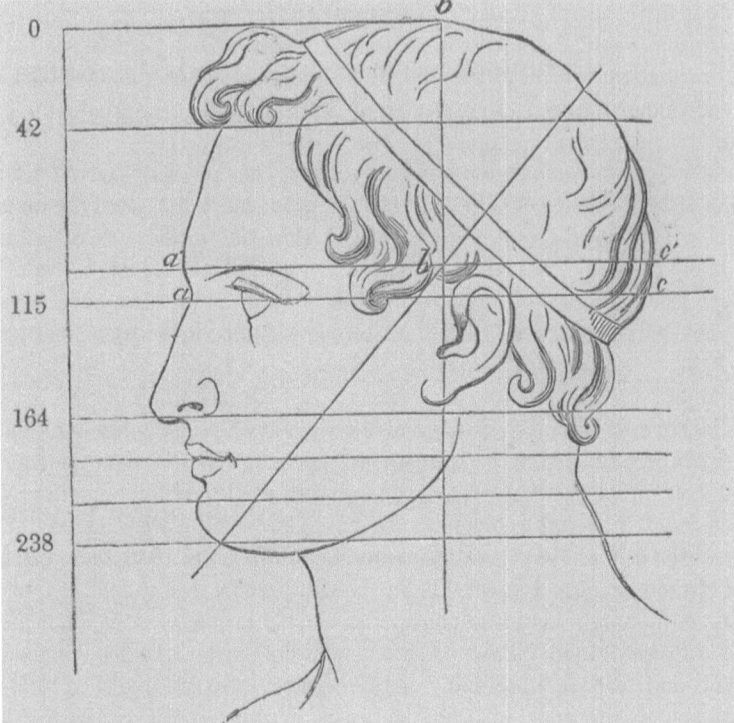

Figure 6.1 The average man.

Source: Quetelet (1870, 206).

measures of central tendency that indicate approximate central positions of a phenomenon are useful ways to simplify the complexity of collected data. There are numerous central tendency measures, with the three main measures being **mean**, **mode**, and **median**, which will be defined and exemplified in the following subsections.

6.2.1 Mean

The mean, or average, of a dataset is calculated by summing all relevant values and dividing by the number of data points (it can be calculated on a Google spreadsheet using `average`). This measure provides a central value around which the data points tend to cluster. For example, to calculate the length of an average Harry Potter book, we sum the word counts of each of the seven official Harry Potter books and divide the total by seven:

$$\frac{76,944+85,141+107,253+190,637+257,045+168,923+198,227}{7}$$

$$=\frac{1,084,170}{7}=154,881.4286$$

Surprisingly, this is almost the same as the average length of a Lord of the Rings book, which is only slightly longer, with 160,367.66 words per book.

Just like comparing J.K. Rowling's Heptalogy to J.R.R. Tolkien's Trilogy, the mean is a useful way of comparing large datasets. For example, research has shown that the cost of labor in England decreased during the 19th century by comparing the average wages of farmworkers in 1832 and 1872 (Clark 2001). The average wage dropped from £29.1 to £21.9, reflecting a general decline in labor costs.

Questions

1. Israeli governments:
   a. *Calculate the average number of ministers in Israeli governments (data is available at https://zefsegal.com/computational-literacy-for-the-humanities/).
   b. *How does the resultant average compare to the first governments of Israel and the latter governments.

6.2.2 Mode

The mean is not always the most useful measure of central tendency. Sometimes it is more convenient to characterize a set of data by the value that appears most frequently, which is called the **mode** (it can be calculated on a Google spreadsheet using mode). For example, when describing countries, we often focus on the largest cities as representative of society as a whole. These cities are the mode since their populations are larger than those of any other settlements. Accordingly, New York represents the United States and Buenos Aires represents Argentina, even though there are often significant social and cultural differences between the largest city and other parts of their respective countries.

The mode is a convenient central measure when it appears much more frequently than other values, but it becomes useless when the number of repetitions is relatively small. For example, in the English town of Accrington in 1830, people died at the ages of 8, 20, 35, 32, 17, 82, 0, 0, and 22 (Canning, 16).

Questions

2. Deaths in Accrington:
   a. *What is the mode of age at death in the town of Accrington?
   b. Is the mode a good representative of the age at death in the town of Accrington?

It is difficult to estimate the importance of this data given the available information. The sample size is very small, and the fact that two out of nine deaths were infants does not indicate a particular trend among the town's population.

6.2.3 Median

The **median** is the middle value in a dataset when the values are arranged in ascending or descending order (it can be calculated on a Google spreadsheet using `median`). It is a slightly more complex central measure than the mode because, unlike the mode, it requires consideration of the central value's position relative to all other values. In the process of calculating the median, the other data points' actual values are ignored, and only their order is considered. If the dataset has an odd number of observations, the median is the single middle value. If the dataset has an even number of observations, the median is the average of the two middle values.

Let us consider the ages of the deceased in Accrington: 8, 20, 35, 32, 17, 82, 0, 0, and 22, then arrange them from smallest to largest. Our new number sequence is 0, 0, 8, 17, 20, 22, 32, 35, 82.

The median is the value that appears in the middle of the list. In this case, the median age at death in the town of Accrington in 1830 was 20, as there are four values smaller than 20 and four values larger than 20. Unlike the mode, which does not describe anything other than itself, the median provides information about the general behavior. Based on the fact that the median age at death is 20, it can be said with certainty that half of the deceased were children and teenagers.

If we add the value 22 to the previous numbers, we get the following ordered list: 0, 0, 8, 17, 20, 22, 22, 32, 35, 82. In this case, there is no single middle value, since 20 and 22 are both in the middle. The median is defined as the average of these middle value, which is 21.

Questions

3. *Calculate the median number of ministers in Israeli governments (data is available at https://zefsegal.com/computational-literacy-for-the-humanities/).

6.2.4 Comparative analysis of central tendency measures

Each of the three central tendency measures offers a different way to summarize a set of data. Choosing between them depends on the type of data we have (e.g., continuous or discrete data), the presence of outliers that may skew measures like the mean, and primarily what we are trying to discover in the data at our disposal.

The **mode** is the simplest measure of central tendency to calculate, but it is not used as often because it only identifies the most frequent value and does not consider its relation to other values. It is most useful when this value is much more common than others and when dealing with categorical data. For example, societies are often described by their predominant religion – the one most people follow. In this case, we cannot calculate an average or median religion, so the mode is the best way to summarize the data.

However, the mode can be problematic:

- When the most common value is not very frequent compared to the rest of the data.
- When there are multiple values that appear with the same frequency.
- When the most common value is an outlier and does not reflect the rest of the data.

The main advantage of the **mean** is that it includes all values in its calculation. However, very large or small values can significantly affect the mean, shifting it up or down. For example, the average salary in most societies is often skewed upwards due to the extremely high salaries of a few wealthy individuals, which does not accurately reflect the incomes of the majority. Consider five monthly salaries: 6,000, 8,000, 10,000, 12,000, and 60,000. The average salary among these is 20,000, but this does not faithfully represent the data. Instead, it reflects the influence of the high outlier.

In contrast to the mode and the mean, the **median** is not influenced by outliers. In the previous example, the median salary is 10,000, which better reflects the salaries of at least four out of the five data points.

To illustrate the differences between the three measures, let's look at one case study. Consider 124 deaths that were registered in Blackbourn, England, in 1880. Table 6.2 details the ages at death, revealing insights into the town's demographic structure (Lancashire 2024).

Questions

4. Calculate the mode, median, and mean of age at death in Table 6.2.

To find the mode, we look for the value that appears most frequently. In this case, the most common age is 0, as 46 out of 124 people died in their first year of life. To

Table 6.2 Age at death in Blackbourne, England, 1880

Age at death	Number	Age at death	Number	Age at death	Number
0	46	27	1	53	2
1	14	28	2	55	1
2	7	31	1	60	3
3	4	32	1	63	1
5	1	34	1	64	1
6	1	35	1	65	4
15	2	36	1	68	1
17	2	41	1	70	3
18	1	43	2	73	2
20	1	45	2	74	1
22	1	47	2	77	1
23	1	49	1	85	1
24	1	51	1	87	1
25	1	52	1	91	1

find the median, we arrange the values in ascending order. The median is the value that has an equal number of values above and below it. With 124 people, the mid-values are in the 62nd and 63rd positions. Both of these values are 2, so the median, which is the average of these two, is 2. To calculate the mean, we sum all 124 values and divide by 124. After doing so, we find that the average age is 20.38.

The mode highlights a harsh reality – the most common age at death was before the individual turned one year old. The median strengthens this observation, indicating that half of the deceased were infants. Nonetheless, on average, people in the town died at age 20.38, a figure that reflects the impact of outliers who lived into their eighties and nineties.

The data from Blackbourn reflects a statistical bias resulting from infant mortality. Societies suffering from high rates of infant mortality, like many societies in the Global South today, have very low life expectancy measures (both median and mean).

Questions

5. Calculate the mode, median, and mean, excluding individuals who died at age 3 or younger.

The answers you obtained suggest a different interpretation of the demographic structure of the English town. The mode is 65, the median is 53, and the mean is 46.9. This indicates that the town is neither a society of elderly people nor a society of children and teenagers. Those who survived infant diseases lived, on average, until their late forties and early fifties.

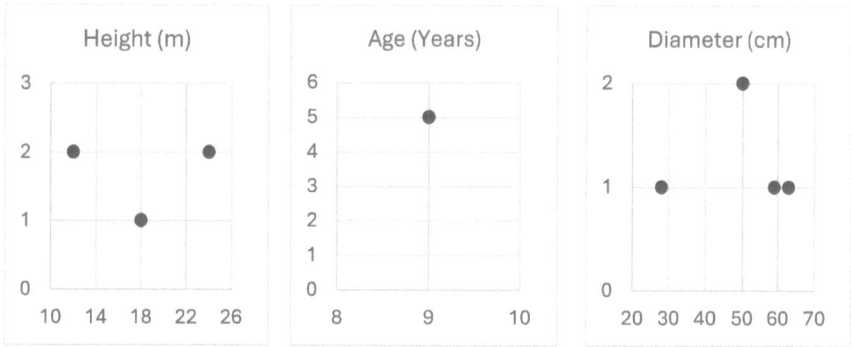

Figure 6.2 Three scatter plots showing the frequency of each of the three data sets.

6.3 Measures of dispersion

Means, medians, and modes summarize data by providing a single value, but it doesn't capture the spread of the data. For example, suppose five different trees in a park are measured for their heights, trunk diameters, and ages, and the results are as follows:

Category	Measurements	Mean	Median
Height (m)	12, 18, 12, 24, 24	18	18
Diameter (cm)	28, 50, 59, 50, 63	50	50
Age (years)	9, 9, 9, 9, 9	9	9

Although the means and medians are identical in each category, a closer look shows that the dispersion of the data is entirely different for each measurement. To see this, we can present the data in three scatter plots (Figure 6.2).

In contrast to the ages of the five trees, which are all equal, the diameters and heights of the trees are spread across a range of values. Measures of central tendency do not convey how much the data values differ from each other or from the central point. To capture this variability, we need to use measures of dispersion. In this context, we will introduce variance and standard deviation.

6.3.1 Variance and standard deviation

Variance, a key measure of data dispersion, was first conceptualized by British mathematician Ronald Fisher in his 1918 paper on genetics. Mathematically, variance measures how much the data points in a dataset differ from the mean. It does this by calculating the average of the squared differences between each data point and the mean (it can be calculated on a Google spreadsheet using `Var.p`).

Let's illustrate this with the trees' measurements. For the tree heights, the measurements are 12, 18, 12, 24, and 24 meters. The mean height is 18 meters. Accordingly, the distance of the first tree from the mean (18) is 6 meters, the distance of the second tree is 0 meters, the distance of the third tree is 6 meters, and the distances of the last two trees are both 6 meters. The farther a data point is from the mean, the greater the deviation.

We square these distances to get the values: 36, 0, 36, 36, and 36, and calculate the average of these five values to find the variance: $\dfrac{36+0+36+36+36}{5} = 28.8$.

Questions

6. *Calculate the variances of the trunk diameters and ages.

The value of the variance is always non-negative (positive or zero) because it is a sum of squares, and the smaller the variance, the more concentrated the dispersion is around the mean. This characteristic allows us to compare the spread of different datasets. However, the resulting numbers can seem arbitrary. For instance, what is the meaning of the value 28.8 we obtained for the heights, or the value 146.8 we obtained for the diameters?

In order to answer that question, we need to define an additional measure: **standard deviation**, which is the square root of the variance (it can be calculated on a Google spreadsheet using `stdev.p`). Since variance measures the average of squared distances from the mean, the standard deviation is approximately the average distance of the data points from the mean. This makes the standard deviation easier to interpret, as it is in the same units as the original data.

Accordingly, the standard deviation of the tree heights is $\sqrt{28.8} \approx 5.37$, the standard deviation of the tree diameters is $\sqrt{148.8} \approx 12.12$, and the standard deviation of the tree ages is 0.

Unlike variance, the standard deviation has an immediate meaning relative to the data, representing the average distance from the mean. The units of measurement for standard deviation are always the same as the units of the data itself. In our example, the standard deviations of heights, diameters, and ages are in meters, centimeters, and years, respectively. Since the standard deviation approximates the average distance of the mean, we can estimate that approximately half of the data points lie within one standard deviation from the mean.

A more precise approximation was established by Russian mathematician Pafnuty Chebyshev (1821–1894), who proved that in any random sample, at least half of the data points will lie within $\sqrt{2}$ standard deviations of the mean. Therefore, by using the mean (a measure of central tendency) in conjunction with the standard deviation (a measure of dispersion), we can more accurately estimate the "significant" range of the data.

For example, the average diameter of the five trees was 50 centimeters, and the standard deviation was approximately 12.12 centimeters. Therefore, most diameters would lie between 50-12.12=37.88 centimeters and 50+12.12=62.12. Indeed, the diameters approximately within this range are 50, 50, and 63 centimeters. We obtained three out of five data points, and we are extremely close to another value, 63 centimeters. Similarly, the average age of the five trees was 9 years, and the standard deviation was 0. All ages are exactly at a distance of 0 from this mean.

The average height of the five trees was 18 meters and the standard deviation was approximately 5.37 meters. Only one value lies between 18-5.37=12.63 meters and 18 + 5.37 = 23.37. However, this range is very close to the actual range of values, which is between 12 and 24 meters, and using Chebyshev's approximation of √2 standard deviation (1.4*5.37 = 7.52) would include the whole data set.

The mean alone does not provide information about the dispersion of the data. The standard deviation alone gives information about the dispersion, but it is always relative to a given mean. Therefore, only the combination of the two allows us to distinguish between "unusual" cases and the "usual" range.

Figure 6.3 depicts the distribution of Body Mass Index (BMI) for four groups: Norwegian youth aged 15–19 in the years 1963–1975, Norwegian adults aged 20–24 in the same years, cadets aged 15–19 at the US Military Academy West Point in the years 1874–1894, and cadets aged 20–24 in the same years. In which cases is the mean higher, and in which cases is the standard deviation higher?

Due to the symmetric shape of each distribution (bell-shaped), it is relatively easy to estimate the mean in each case, as it would be approximately at the peak

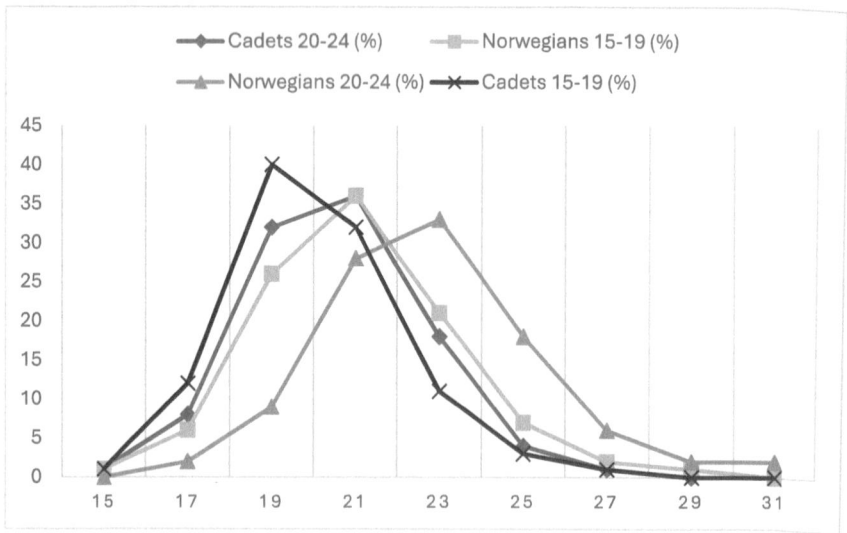

Figure 6.3 BMI distribution by age and sample group.

Source: Based on data from Cuff (1993).

of the bell curve. The average BMI of the young Norwegians is approximately 21, while the average BMI of the adult Norwegians is approximately 23. In comparison, the average BMI of the cadets is about two units less for each age group, with 19 for the teenagers and 21 for the adults. The difference between the means likely reflects the fitness level of the cadets at the military academy at the end of the 19th century.

Standard deviation measures the dispersion and is visually represented by the width of the different "bells". Since the four curves look almost identical except for the different means, we can estimate that the standard deviations are also nearly identical. The reason for the similarity between these four distributions will be explained in the next section.

6.3.2 Normal distribution

In statistics, a distribution describes the probabilities of different outcomes for a given phenomenon or process. Among the many possible distributions, the **normal distribution** is particularly significant due to its frequent occurrence in natural and social sciences.

In 1846, the Belgian statistician Quetelet undertook a study where he collected the chest measurements of 5,738 Scottish soldiers (Quetelet 1846). His results are presented in Table 6.3.

Looking at the histogram representation of Quetelet's data (Figure 6.4), we observe that only a few soldiers had a chest circumference of 33 inches (about 84 cm), indicating a slim build. Similarly, only a few soldiers had an impressive chest circumference of 48 inches (about 122 cm). The similarity between the number of soldiers with narrow chests and those with wide chests is striking.

Table 6.3 Chest measurements of Scottish soldiers (in inches)

Chest circumference	Frequency
33	3
34	18
35	81
36	185
37	420
38	749
39	1073
40	1079
41	934
42	658
43	370
44	92
45	50
46	21
47	4
48	1

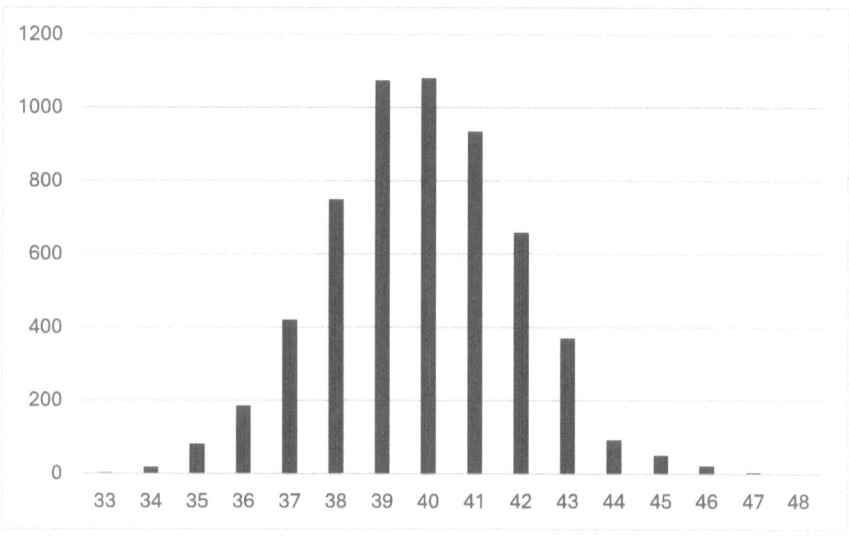

Figure 6.4 A histogram representation of the data in Table 6.3.

The distribution of chest circumferences is symmetric, with a large concentration around the center of the histogram (39–40 inches) and a gradual decrease on both sides, nearly vanishing at the edges.

This type of distribution, resembling a bell curve, is common to many phenomena, particularly in complex biological and social contexts. Due to its typical and frequent occurrence, it was named the "normal distribution" by two 19th-century scientists, Charles S. Peirce (1873) and Francis Galton (1877).

It is important to understand that real-life examples of normal distributions, such as the distribution of heights, weights, or test scores, are only approximations. A true normal distribution, as depicted in theoretical models, is a mathematical abstraction and not an exact representation of any real phenomenon.

The normal distribution is important for both theoretical and practical reasons. First, it is a very common distribution in real-world data. Second, the **central limit theorem**, a fundamental principle in statistics, states that if you take a large enough number of random samples from any population, and you calculate the average (mean) of each sample, the distribution of those sample means will be approximately normal. This becomes more accurate as the sample size increases. This explains why the normal distribution is so prevalent, as many measured values in nature and society are the result of numerous random factors.

Practically, characterizing a distribution as normal is advantageous because it provides a fixed relationship between central tendency measures, dispersion measures, and the behavior of values. In a normal distribution, the mean, median, and mode are identical. Furthermore, 68% of the data falls within one standard deviation of the mean, 95% within two standard deviations, and 99.7% within three standard deviations. This predictable pattern simplifies the analysis and interpretation of data.

Returning to the example of the chest measurements of Scottish soldiers: the mean chest measurement is 39.8 inches, and the standard deviation is 2.05 inches. The range of values within one standard deviation of the mean (37.75 to 41.85 inches) includes 3,835 out of 5,738 soldiers, or 67%. The range of values within two standard deviations of the mean (35.74 to 43.86 inches) includes 5,468 soldiers, or 95.3%. Finally, the range within three standard deviations of the mean (33.65 to 45.95 inches) encompasses 99.5% of all soldiers.

Questions

7. In a study on the heights of men and women in Bavaria in the 19th century, data were collected on 5000 men recruited into the Bavarian army between 1810 and 1840 (Baten and Murray 2000). The mean of these data is 166.8 cm, and the standard deviation is 6.4 cm. Assuming this sample represents the entire Bavarian society, answer the following questions:
    a. *What is the range of heights within one standard deviation of the mean, and what percentage of men fall within this range?
    b. *What range characterizes about 95% of the men in Bavaria?
    c. *What height would we consider as exceptionally tall (since 99.85% of the population is shorter) in 19th-century Bavaria?

6.3.3 Non-normal distributions

Of course, not every set of data follows a normal distribution. For example, income distribution is almost never normal because the number of wealthy individuals is much smaller than the number of poor individuals. This results in a skewed distribution rather than a symmetric bell curve.

The distribution shown in Figure 6.5 is not normal because it has two peaks, neither of which is at the mean. One peak represents the world's poor, earning about one dollar a day, mostly in Asia and Africa. The other peak is due to the world's richest nations, where earnings are about 15–20 dollars a day, primarily in Europe and America. "Non-normal" distributions are significant in their own right because sometimes they highlight inequality issues, as depicted in this diagram.

Although this book primarily focuses on the normal distribution, it is important to understand that most real-world data do not follow a normal distribution. Sometimes, distributions are not normal because the sample size is too small, as in the case of measuring five trees. Other times, the distribution is skewed, as seen in the previous income example. While the concept of the normal distribution is

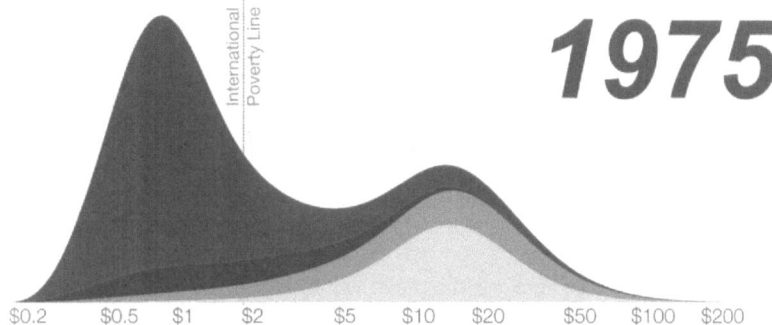

Figure 6.5 Daily income per person in 1975, calculated based on 2011 currency value (Roser 2017). The X-axis represents daily income, and the Y-axis represents the number of people receiving that income. The colors reflect different continents: light gray for Europe, medium gray for North and South America, medium-dark gray for Asia and Oceania, and dark gray for Africa.

useful in relevant cases, in many situations, we rely on the quantitative measures obtained in our calculations: mode, median, mean, and standard deviation.

6.3.4 Skewness

In a normal distribution, the mode, median, and mean are identical, making any of them a valid measure of central tendency. Additionally, an equal number of cases lie above and below the mean. However, this alignment does not hold true for non-normal distributions, such as the global income distribution. In such cases, using the mean can be misleading. For example, in the United States in 2019, the average salary was $51,916, while the median salary was $34,248. The significant gap of about $17,000 between these two measures arises, because the top 1% of earners received about 20% of the total income, while the bottom 10% received only about 1.7% of the total income.

So far, we have discussed two types of measures: central tendency measures and dispersion measures. However, neither of these helps identify asymmetry in a distribution. To address this, we introduce a new concept: **skewness**. Skewness directly measures the degree of asymmetry in a dataset (it can be calculated on a Google spreadsheet using skew.p). This concept is particularly important in analyzing social processes, which often do not follow a normal distribution.

To calculate skewness, perform the following steps:

1. Calculate the cubed distances of each data point from the mean.
2. Sum these values and divide by the product of the total number of observations and the standard deviation cubed.

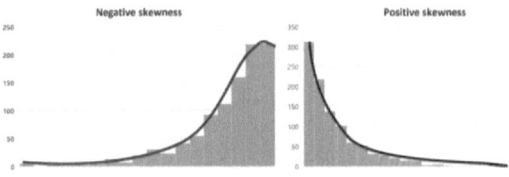
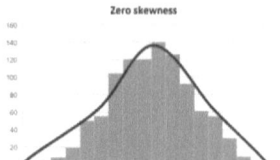

Figure 6.6 Positive, negative, and zero skewness.

The resulting formula (where $\bar{x}$ is the mean value) is:

$$\frac{\Sigma(x-\bar{x})^3}{(\text{number of data points})(\text{standard deviation})^3}$$

Skewness identifies the presence of asymmetry, its magnitude, and its direction. A skewness of zero reflects perfect symmetry, a positive skewness value reflects a distribution where most values are smaller than the mean, and a negative skewness value reflects a distribution where most values are larger than the mean (Figure 6.6). Furthermore, values higher than 1 or less than -1 reflect high skewness (significant bias in a particular direction). Values between -0.5 and 0.5 reflect low skewness.

Returning to the measurements of the trees in the park, we previously calculated that the mean height was 18 meters, with a standard deviation of 5.37 meters. Now, we will calculate the skewness of the tree heights. For the heights of 12, 18, and 24 meters, we subtract the mean and cube the result:

- For a height of 12 meters: $12-18 = (-6)$, and $(-6)^3 = -18$.
- For a height of 18 meters: $18-18 = 0$, and $0^3 = 0$.
- For a height of 24 meters: $24-18 = 6$, and $6^3 = 18$.

Since the values 12 and 24 each appeared twice, we multiply the results by two. The skewness is then calculated as: $\dfrac{2 \cdot (-216) + 0 + 2 \cdot (216)}{5 \cdot (5.37)^3} = 0$

This result indicates that the distribution of the tree heights is perfectly symmetrical with respect to the mean, showing no skewness.

Questions

8. *Calculate the skewness of the diameters and the ages of the trees.

The skewness of the ages is zero since there is a single value. The skewness of the diameters is moderately negative (between -1 and -0.5) and reflects a slight bias towards the larger values. We can infer that the outlier of this distribution is the lower value of 28.

Looking at the larger dataset from 1880 Blackbourn, we found that the average age at death was 20.38 years. Calculations show that the standard deviation was 18.98 years, and the skewness was 2.907. Based on the standard deviation, we can identify a relatively large range of "normal" values between approximately 1 year to 39 years at death. However, the high positive skewness reflects a large bias towards the younger ages. We can infer that the range of normalcy is not symmetric and is heavily skewed. Ages on the older side of our range were actually outliers, let alone, those that were outside this range, such as those who passed away in their eighties and nineties.

6.4 Case study: Distant reading of books by Wharton and Stein

At the end of Chapter 5, we introduced the concept of "distant reading", a research method developed by the Italian-American scholar Franco Moretti. In distant reading, computational and statistical tools are used to create an abstraction of a text or, more often, a selection of texts, without actually reading them "closely". We will end this chapter with an exercise in which we will conduct a distant reading of four books: two novels and two autobiographies by two American authors with similar biographical profiles: Edith Wharton and Gertrude Stein.

Edith Wharton (1862–1937) was born in New York. She wrote over 40 books, including novels, short stories, poetry, and non-fiction. In addition, she was a highly regarded landscape architect and interior designer. In 1921, she became the first woman to win the Pulitzer Prize for her novel *The Age of Innocence*, which was also adapted into a film directed by Martin Scorsese in 1993. In 1907, Wharton moved to France, where she continued to write until her death in 1937.

The novel we will explore is Wharton's *The House of Mirth* (1905). In this novel, Wharton depicted the stratified society in which she grew up and its response to social changes at the turn of the century.

Wharton's autobiography *A Backward Glance* was published in 1934. In it, she described her private and public life, the aristocratic society of New York, her many travels around the world, and her achievements as a writer.

Gertrude Stein (1874–1946) was born in Pittsburgh and grew up in Oakland, California. Stein is known as a writer, poet, playwright, art collector, and leading figure in the modernist movement in literature and art. Like Wharton, Stein made her home in France, to where she moved in 1903 and where she passed away in 1946.

The novel we will explore is *Three Lives*, which was her first and most famous novel. It was published in 1905–1906 after Stein had difficulty finding a publisher due to the unconventional style of the book. The novel consists of three novellas set in a fictional town.

Stein's autobiography *The Autobiography of Alice B. Toklas* was published in 1933. The autobiography is told ostensibly through the eyes of her partner, narrating their life together.

6.4.1 Data collection

An excellent source for digital versions of books is the Project Gutenberg website (www.gutenberg.org/). This is an initiative to create a computerized library of books and other publications, primarily in English. It was founded in 1971 by Michael Hart, a student at the University of Illinois, who started it by typing the United States Declaration of Independence into the university's computer system and continued with the works of Shakespeare and the Bible. Since then, thousands of volunteers have joined the project, scanning and typing over 70,000 titles, as of February 2024.

All the texts in the project are in the public domain according to copyright laws. The books can be read on a computer screen or downloaded as files in various formats: text files (txt), PDF, HTML, and sometimes even as audio files (MP3). We will analyze the textual format (txt) of the four mentioned books, which can be found at https://zefsegal.com/computational-literacy-for-the-hum anities/.[1]

For each of the four books, we will collect three types of data: (1) words and their frequency, (2) word lengths, and (3) sentence lengths. Most of the operations we need have been described in previous chapters, and a few will be new. More importantly, we will save these data as CSV and text files that we will use later in the chapter.

Words and their frequency

Questions

9. Create word frequency lists:
   a. Prepare the two novels and the two autobiographies, by reading text files (see section 3.8), normalizing them (see section 3.7) and creating token lists (see solution for Question 16, Chapter 5)
   b. Sort the resulting frequency lists (see section 5.3.2).
   c. Create output files with the frequency lists for each book, save them under the titles: "WordsWhartonNovel.csv", "WordsWhartonBio.csv", "WordsSteinNovel.csv", "WordsSteinBio.csv" (see sections 3.8.3 and 5.3.3).
   d. Import the four CSV files into Google Sheets. When prompted, select "Insert new sheet(s)" instead of "Create new spreadsheets".

Each of the first four spreadsheets is structured into two columns: the first column lists the frequency of each word type, while the second column contains the words themselves.

We will use these spreadsheets to calculate the following data for each book, using the functions listed in parentheses:

- The number of words (`sum`), i.e., tokens
- Number of unique words (`count`), i.e., types
- Type–token ratio, i.e., types/tokens
- Standard Deviation (`stdev.p`)
- Mode (`mode`)
- Median (`median`)

For example, the median of word frequencies can be calculated using the formula =`median(A:A)`. The software will perform the calculation on all values in column A without you needing to specify the exact range of values (i.e., =`median(A1:A12568)` for Wharton's biography).

Questions

10. Calculate the six statistical measures for all four books. After calculating the data for the first book, you can copy the formulas and paste them into the three remaining sheets. The calculations will update automatically for each sheet. (The answer appears in Table 6.5.)

Type–token ratio

In Chapters 3 and 5, we discussed type–token ratio (TTR) as an indicator of lexical richness. The inverse of this measure, namely the number of tokens divided by the number of types, produces the **average frequency** of words in the text. The average word frequencies for our four texts are as follows: Wharton's autobiography has an average frequency of 8.255, while her novel has a frequency of 10.469. Stein's autobiography shows an average word frequency of 13.602, and her novel has the highest frequency, at 25.873.

What is the difference between TTR and average frequency? Why is average frequency not a valid measure for comparison between texts?

When comparing different datasets, relying on absolute values such as frequencies can be misleading if we aim to compare anything beyond their sheer size. For instance, consider the comparison of the number of women in the American Congress versus the Rwandan Parliament. As of 2024, the US Congress has 151 female members, while the Rwandan Parliament has 58 female members. This difference largely stems from the disparity in the total number of members: the US

Congress (Senate and House of Representatives) comprises 535 members, whereas the Rwandan Parliament (Senate and Chamber of Deputies) has only 106 members.

To make a meaningful comparison, we use relative frequency, also known as **normalized frequency**. This metric is calculated by dividing the relevant frequency by the total number of members. For example, the normalized representation of women in the American Congress is 151/535, which equals 0.28, or 28.2%. This means that for every 100 members of congress, 28 of them are women. In contrast, the normalized representation of women in the Rwandan Parliament is 58/106, which equals 54.7%. This comparison reveals that the representation of women in Rwandan politics is significantly higher than in the US (54 in 100 compared to 28 in 100).

Similarly, when comparing lexical richness of texts, it is essential to account for the size of the text (i.e., the number of tokens). The average frequency of words in one text may be lower than the average of another text, yet the second text may exhibit relatively less word repetitions when the lengths of the texts are considered. Unlike the average frequency, which is an absolute value, the TTR measures the proportion of unique words relative to the size of the text, with a higher proportion indicating greater lexical richness. However, as mentioned in section 5.3.5, this measure is not immune to variations in text length, as longer texts tend to yield lower ratios due to the natural repetition of words. Nevertheless, considering the relatively similar lengths of our four books, the TTR will suffice.

Questions

11. *Determine whether other statistical measures of word frequency are influenced by the length of the texts.

Word lengths

Word length is an additional measure that allows comparison between texts. In question 9, in the process of creating a word frequency list, we created a list of tokens. We will now iterate through this list and create a parallel one, `length_l`, that contains the lengths of the words. For each index i, the length of `tokens[i]` is stored in `length_l[i]`.

```
length_l = []
for word in tokens:
    length_l.append(len(word))
```

To calculate statistical measures regarding word lengths in the text, we will use functions from the `statistics` library.

Libraries in Python are collections of pre-written code that you can use in your Python programs to perform specific tasks without having to write all the code yourself. Libraries typically contain modules or packages that include functions that help you achieve particular functionalities more easily.

To use the various tools of the `statistics` library, we need to **import** it. Once the library is imported, functions can be called with the name of the library prefixed to their name, separated with a period. The four statistical functions in the following code take as input a list of numeric values and return the appropriate measure. We will call them here with `length_l` as an argument, to calculate the various statistical attributes of the list of word lengths.

```
import statistics
mean = statistics.mean(length_l)
std = statistics.pstdev(length_l)
mod = statistics.mode(length_l)
med = statistics.median(length_l)
```

We will save the results in a CSV format (comma-separated values) where each measure name and its value are separated by a comma.

```
Mean,4.5734498
...
```

Instead of saving the results of each book separately, we will save the data of all the books into a single file titled "WordLengths.txt". Book entries will be separated by a dashed line, '-----'.

Since we will run this program four times, once for each book, we want to distinguish between the first iteration, in which we will create the file, and three iterations, in which we append to an existing file. Remember that when we open a file for writing with the argument `w`, if the file exists, the opening process overwrites it and creates a new file. This is what we will do for the first book. Subsequently, to add information to the existing file, we will use the `a` argument (from the word append). In the following code, we ask the user whether to append the result.

```
reply = input("Do you want to append the results to
an existing file? (y/n)\n")
if reply == "y":
    output_file = open("WordLengths.txt", "a")
    output_file.write("-----\n")
```

```
else:
    output_file = open("WordLengths.txt", "w")
```

The following code adds five lines to "WordLengths.txt". The first line writes the title of the book, which we will denote by the variable `filename`, which was used in opening the respective text file, and the consequent lines include the four statistical measures.

```
output_file.write("File," + filename + "\n")
output_file.write("Mean," + str(mean) + "\n")
output_file.write("Standard  deviation,"  +  str(std)
+ "\n")
output_file.write("Mode," + str(mod) + "\n")
output_file.write("Median," + str(med) + "\n")
output_file.close()
```

At this point, our five datafiles – "WordsWhartonNovel.txt', "WordsWhartonBio. txt", "WordsSteinNovel.txt", "WordsSteinBio.txt", and "WordLengths.txt" – contain frequency lists and statistical data regarding the length of words. We will now turn to the lengths of the sentences in each book.

Sentence lengths

In Chapter 3, we measured the average length of sentences in two stories – "The Haunted Mind" by Nathaniel Hawthorn and "Two Kinds" by Amy Tan. To identify sentences, we counted the number of periods, which gave us a rough estimate of the number of sentences in the text. However, not every sentence ends with a period, and not every period marks the end of a sentence. To achieve a more accurate count, we need to address specific exceptions, such as removing periods following abbreviations like 'Mr.' and 'Mrs.' and treating question marks as sentence-ending punctuations. While you now have the ability to write a program that performs these text alterations, it is challenging, if not impossible, to anticipate all possible scenarios. For example, consider the sentence: "Jim B. Jones has completed his Ph.D. degree in the Dept. of Computer Science!" Counting all the periods as sentence ends would incorrectly yield four sentences and miss the actual sentence that ends with an exclamation mark.

Better sentence tokenization can be performed using built-in functions from the `nltk` (Natural Language Toolkit) library. For our particular purpose, the `nltk` library provides the function `sent_tokenize()`, which takes a text (string) as an argument, and returns a list of sentences. This function is found in the `punkt` module of the `nltk` library, which includes functions and methods used for processing texts. The function addresses the challenge of identifying sentences using a machine learning approach.

To use the various tools of the `nltk` library, we need to import the library and download the relevant module. These actions are simple and performed using two Python commands:

```
import nltk
nltk.download("punkt")
```

The following program prepares a text for analysis and creates a list of all the sentences. It assumes that the content of the text file was previously read and assigned to the variable `rawtext`.

```
# Replace newline characters and space before periods
cleantext  =  rawtext.replace("\n",  "  ").replace
(" .", ".")
# Tokenize the text into sentences
sentences = nltk.sent_tokenize(cleantext)
```

Questions

12. Follow the instructions given in subsection 6.4.2 "Word Length Frequency":
    a. *Create a list of sentence lengths (in terms of the number of words) and assign it to the variable `length_l`.
    b. *Calculate all the relevant statistical measures of this list using the statistical functions provided in the statistics library.
    c. *Save the results to a text file "SentenceLengths.txt", which will eventually contain the results for all four books.
13. *Are any of the statistical measures of word length or sentence length dependent on the text length?

6.4.2 Distant reading

We chose to conduct a distant reading of four literary works from two genres written by two authors with similar biographical profiles. With this particular corpus, we can analyze the books according to two dimensions: author and genre.

Based on the data we collected (see Table 6.5), we will compare the books, authors, and genres.

- What can be learned about the books and authors from statistical data?
- Is there more similarity between books written by the same author or between books of the same genre?

The data highlight differences between these books, but also reveal some common features. Two features, in particular, are common to all four books, and indeed to English texts in general:

- **Word Frequency**: The shared mode (1) and similar median (1 and 2) are due to the known distribution of word frequencies in texts overall. As seen in Chapter 5, Zipf's Law applies to texts of various types and lengths. In long texts, 40% to 60% of words are hapaxes, so the mode is usually 1. The low median also results from the abundance of unique words in texts.
- **Word Length**: The most common word length in all texts is 3. Reviewing the word lists created for each text, we see that among the most frequent words in all texts are many three-letter words. Some are common in any English text (e.g., 'and', 'the', 'was'), while others are specific to the text's content (e.g., 'her', 'she').

However, the differences between the texts outweigh their common features:

- **TTR**: Wharton's books have a much higher type–token ratio (TTR) than Stein's books. Additionally, the TTR of both autobiographies is higher than that of their respective novels, although the difference between Wharton's autobiography and novel is less pronounced.
- **Average Sentence Length**: The average length in the two novels is almost identical – about 21 words. However, there is a significant difference in the autobiographies: while Stein tends to use sentences of similar length to those in her

Table 6.4 Our corpus in a 2x2 dimensional model

	Novel	*Autobiography*
Wharton	The House of Mirth	A Backward Glance
Stein	Three Lives	The Autobiography of Alice B. Toklas

Table 6.5 A summary of the statistical measures for all four books

	Wharton's bio	Wharton's novel	Stein's bio	Stein's novel
Word Frequency				
Tokens	103,745	129,099	92,380	86,052
Types	12,387	11,821	6,772	3,323
TTR	0.119	0.092	0.073	0.039
Stand. Dev.	94.9	117.9	109.5	144.3
Mode	1	1	1	1
Median	1	1	2	2
Word Length				
Average	4.57	4.47	4.31	4.05
Stand. Dev.	2.59	2.45	2.27	1.9
Mode	3	3	3	3
Median	4	4	4	4
Sentence Length				
Average	32.76	22.28	18.93	21.08
Stand. Dev.	20.98	15.02	12.38	15.35
Mode	23	10	11	10
Median	30	19	16	17

novel, the sentences in Wharton's autobiography are much longer – about 33 words. Additionally, half of the sentences in that book are at least 30 words long.
- **Average Word Length**: The average word length in all four books is similar (between 4 and 4.5), although it is slightly lower in Stein's books. However, the standard deviation of average word length in Wharton's books is higher in both the novel and the autobiography, indicating a wider range of lengths.

These differences point to a significant stylistic divergence between Wharton and Stein. Wharton's language is richer, with a more varied vocabulary and longer sentences. Her books are also longer. Stein's language, on the other hand, is characterized by a simpler vocabulary and more repetition. Her sentences are shorter, especially in her autobiography.

We will not delve deeply into the writing styles of the two authors in this book, but we will present two representative quotes from literary criticism that illustrate their distinct writing styles.

In the introduction to a new edition of "Paris France," Stein's 1940 book, Adam Gopnik (2013) wrote:

It is the most deliberately naïve style in which any good writer has ever worked, and it is also the most 'faux-naïf', the most willed instance of simplicity rising from someone in no way simple … Stein's style is to writing what sushi is to cooking – not so much an example as a repudiation of the whole idea that still manages to serve the original function.

In the introduction to a collection of early critical essays on Wharton's writings, it is noted that the author faced sharp criticism early in her career for writing about the elite class of New York and her detachment from American society. However,

> Mrs. Wharton was simultaneously recognized as a writer of exceptional literary distinctions ... The key words and phrases that stand out in these [early] reviews are illuminating. She was praised for her 'clarity', ... 'skillful, finished writing', 'command of good English', and 'mastery of language' ... Her work was, in short, 'serious art'. (Tuttleton, Lauer, and Murray 1992, ix)

As we can see, the insights gained through distant reading align with the characterization of Wharton's prose as realistic and "high", while Stein's writing is viewed as modernist.

Glossary

Bell Curve Another term for the normal distribution, referring to its characteristic shape in which most observations cluster around the central peak and taper off symmetrically towards the tails.

Central Tendency A statistical measure used to determine a single score that defines the center of a distribution. The three main types are **mean**, **median**, and **mode**.

Dispersion A measure of how spread out the values in a dataset are, commonly represented by **variance** and **standard deviation**.

Distant Reading A method of analyzing texts by using computational tools to identify patterns and trends in large data sets, without close reading of the texts themselves.

Frequency The number of times a particular value appears in a dataset.

Lexical Richness The diversity of vocabulary used in a text, often measured by comparing the number of unique words to the total number of words.

Mean The arithmetic average of a set of numbers, calculated by summing all the values and dividing by the number of values.

Median The middle value in a data set when the numbers are arranged in ascending or descending order. If there is an even number of values, the median is the average of the two middle values.

Mode The value that appears most frequently in a data set.

Normal Distribution A bell-shaped probability distribution that is symmetric about the mean. It is a common distribution for continuous variables in many natural and social phenomena.

Outlier An observation point that is distant from other observations in a data set. Outliers can heavily influence statistical measures like the mean and variance.

Sample A subset of a population used to represent the entire population in statistical analysis.

Skewness A measure of the asymmetry of the probability distribution of a real-valued random variable. Positive skewness indicates a longer tail on the right side of the distribution, and negative skewness indicates a longer tail on the left.

Standard Deviation The square root of the variance, representing the average amount by which each data point differs from the mean.

Type–token ratio (TTR) A measure of lexical richness in a text, calculated by dividing the number of unique words (types) by the total number of words (tokens).

Variance A measure of the spread between numbers in a data set, calculated by averaging the squared differences from the mean.

Note

1 For consistency, all four files were downloaded with UTF-8 encoding. In addition, these files were "cleaned" by erasing general information about Project Guttenberg.

References

Baten, Jörg, and John E. Murray. 2000. "Heights of Men and Women in 19th-Century Bavaria: Economic, Nutritional, and Disease Influences". *Explorations in Economic History* 37(4): 351–369. https://doi.org/10.1006/exeh.2000.0743

Canning, John. 2014. *Statistics for the Humanities*. http://statisticsforhumanities.net/book/

Clark, Gregory. 2001. "Farm Wages and Living Standards in the Industrial Revolution: England, 1670–1869". *Economic History Review* 54(3): 477–505.

Cuff, Timothy. 1993. "The Body Mass Index Values of Mid-Nineteenth-Century West Point Cadets". *Historical Methods: A Journal of Quantitative and Interdisciplinary History* 26(4): 171–182. https://doi.org/10.1080/01615440.1993.9956353

de Morgan, Augustus. 1851/1882. Letter to Rev. Heald 18/08/1851. In *Memoirs of Augustus de Morgan by his wife Sophia Elizabeth de Morgan with Selections from his Letters*, edited by Augustus De Morgan, Sophia Elizabeth.

Gopnik, Adam. 2013. "Introduction". In *Paris, France*, edited by Gertrude Stein. Liveright.

Lancashire BMD: Births, Marriages and Deaths on the Internet. 2024. September 20. www.lancashirebmd.org.uk/birthsearch.php.

Mendenhall, Thomas Corwin. 1887. "The Characteristic Curves of Composition". *Science* 9(214): 237–246. https://doi.org/10.1126/science.ns-9.214S.237

Pew Research Center. 2020. "What Census Calls Us". February 6. www.pewresearch.org/social-trends/feature/what-census-calls-us/.

Quetelet, Adolphe. 1846. *Lettres sur la théorie des probabilités, appliquée aux sciences morales et politiques*. M. Hayez.

Quetelet, Adolphe. 1870. *Anthropométrie, ou Mesure des différentes facultés de l'homme*. C. Muquardt.

Riley, James C. 2005. "Estimates of Regional and Global Life Expectancy, 1800–2001". *Population and Development Review* 31(3): 537–543. https://doi.org/10.1111/j.1728-4457.2005.00083.x

Roser, Max. 2017. "The History of Global Economic Inequality". https://ourworldindata.org/the-history-of-global-economic-inequality.

Spitzer, Yannay. 2021. "Pogroms, Networks, and Migration: The Jewish Migration from the Russian Empire to the United States 1881–1914". *Discussion Paper* 21.03. Maurice Falk Institute for Economic Research in Israel. https://en.falk.huji.ac.il/book/2020-2

Tolchin, Martic. 1965. "Births Up 9 Months After the Blackout". *N. Y. Times*, August 10.

Tuttleton, James W., Kristin O. Lauer, and Margaret P. Murray. 1992. "Introduction". In *Edith Warton: The contemporary Reviews*, edited by James W. Tuttleton, Kristin O. Lauer, and Margaret P. Murray. Cambridge University Press.

Udry, J. Richard. 1970. "The Effect of the Great Blackout of 1965 on Births in New York City". *Demography* 7(3): 325–327.

Solutions to selected questions

1. Israeli governments
   a. 21.08.
   b. The average number of ministers in Israeli governments from 1948 to 2024 has nearly doubled compared to the early governments and is almost half of the number seen in more recent governments. This demonstrates a clear increase in the size of the government over time. The Israel Democracy Institute criticized this historical trend and claimed in a report from May 2019 that this growth was a result of political demands required to maintain the coalition rather than efficiency and governmentality.

2. Deaths in Accrington
   a. 0.

3. 21

6. The variances are as follows:

$$\frac{(28-50)^2+(50-50)^2+(59-50)^2+(50-50)^2+(63-50)^2}{5} = 146.8,$$

$$\frac{(9-9)^2+(9-9)^2+(9-9)^2+(9-9)^2+(9-9)^2}{5} = 0$$

7. People's heights in Bavaria:
   a. The range of heights within one standard deviation of the mean is calculated as follows: 166.8 + 6.4 = 173.2 cm and 166.8 - 6.4 = 160.4 cm. Therefore, the range is 160.4 cm to 173.2 cm. In a normal distribution, 68% of the men fall within one standard deviation of the mean.
   a. 95% of the men fall within two standard deviations of the mean. This range is calculated as follows: 166.8 + (6.4 * 2) = 179.6 cm and 166.8 - (6.4 * 2) = 154 cm. Therefore, the height of about 95% of the men was in the range [154, 179.6] cm.
   b. 99.7% of the population lies within a distance of three standard deviations from the mean. This range is calculated as follows: 166.8 + (6.4 * 3) = 186 cm and 166.8 - (6.4 * 3) = 147.6 cm. Therefore, the heights of 99.7% of the men were within the range [147.6, 186] cm. Due to the symmetry of the bell curve, 0.15% of the men were taller than 186 cm and 0.15% were shorter than 147.6 cm.

8. The skewness of the diameters is -0.868 and the skewness of the ages is 0.
11. Standard deviation measures the dispersion of values around the mean and is inherently dependent on the length of the text. Similarly, both the mode and median are influenced by text length. However, due to the nature of word frequency distributions in texts (as explained by Zipf's Law), most words tend to be hapaxes, regardless of text length. Consequently, the mode and median typically do not provide meaningful comparisons between texts.
12. Creating a list of sentence lengths.
   a. Creating a list

```
length_l = []
for sentence in sentences:
    words = sentence.split()
    length_l.append(len(words))
```

   b. Calculating statistics

```
import statistics
mean = statistics.mean(length_l)
std = statistics.pstdev(length_l)
mod = statistics.mode(length_l)
med = statistics.median(length_l)
```

   c. Saving results

```
reply = input("Do you want to append the results to
an existing file? (y/n)\n")
if reply == "y":
    output_file  =  open(path  +  "SentenceLengths.
    txt", "a")
    output_file.write("-----\n")
else:
    output_file  =  open(path  +  "SentenceLengths.
    txt", "w")
output_file.write("File," + filename + "\n")
output_file.write("Mean," + str(mean) + "\n")
output_file.write("Standard  deviation,"  +  str(std)
+ "\n")
```

```
output_file.write("Mode," + str(mod) + "\n")
output_file.write("Median," + str(med) + "\n")
output_file.close()
```

13. The lengths of words and sentences have very little correlation with the length of the whole text. Measures of dispersion, such as standard deviation, are more sensitive to text length, as they depend on the variability of word and sentence lengths. Longer texts tend to have more variability, which can affect the standard deviation. However, there is no need to normalize these measures.

7   Exploring relationships between variables

Up to this point, we have focused on analyzing statistical phenomena in isolation, examining distributions and central tendencies without considering potential interdependencies. However, it's often important to understand the relationships between two or more variables. For example, is there a correlation between life expectancy and geographic location? Does the spread of European imperialism in the 18th century relate to the expansion of the printing industry? Can statistical patterns, such as word lengths or sentence structures, help identify a literary style? In this chapter, we will explore how to measure the strength and direction of relationships between pairs of quantitative phenomena. Additionally, we will learn how to predict the value of one variable based on the value of another.

When high values of one variable consistently occur alongside high values of another, we describe this as a positive correlation. For example, human height and weight show a positive correlation: taller individuals generally weigh more. Conversely, when high values of one variable align with low values of another, we observe a negative correlation. A classic example is the negative correlation between infant mortality rates and time: as medical advances occur over time (e.g., the year 1821 versus 2021), infant mortality rates decrease. It's important to note that exceptions exist – some tall people weigh less, and some short people weigh more. Similarly, while infant mortality still exists today, the overall trend reflects a decline. Correlation indicates a general positive or negative trend, not a strict rule without exceptions.

Moreover, while correlation indicates a relationship between variables, it does not explain the nature or cause of this connection. It is important to remember that a positive or negative correlation does not imply causality. For example, both height and weight are influenced by various factors, such as genetics and nutrition, but they do not directly cause one another. Correlation only highlights the association between the variables, leaving the reasons behind that association open to further investigation.

The degree of correlation between two variables is quantified by a single value known as the **correlation coefficient**. Similar to the different methods of determining central tendency, there are multiple ways to calculate this coefficient. Typically, the coefficient ranges from -1 to 1: a value of 0 indicates no correlation

DOI: 10.4324/9781003502814-8

(independence) between the variables, +1 represents a perfect positive correlation, and -1 indicates a perfect negative correlation. A perfect correlation implies that one variable's value can be exactly predicted from the other.

7.1 Pearson's correlation coefficient: evaluating linear relationships

The most widely used method for calculating correlation is **Pearson's correlation coefficient**. The method was developed by Karl Pearson (1857–1936) in 1896, however, the concept was loosely introduced about a decade earlier by his teacher, Francis Galton (1822–1911). Pearson's correlation coefficient quantifies the relationship between two quantitative variables by comparing the scatter plot of the data to a straight line. This method captures the strength and direction of the relationship, summarizing it with a single numerical value (Figure 7.1).

In the top-left and bottom-left charts, the points are perfectly aligned along a straight line, showing positive and negative trends, respectively. These result in Pearson's correlation coefficient of 1 and -1. In the other charts in the top and bottom rows, the points show clear trends but are not perfectly aligned, yielding

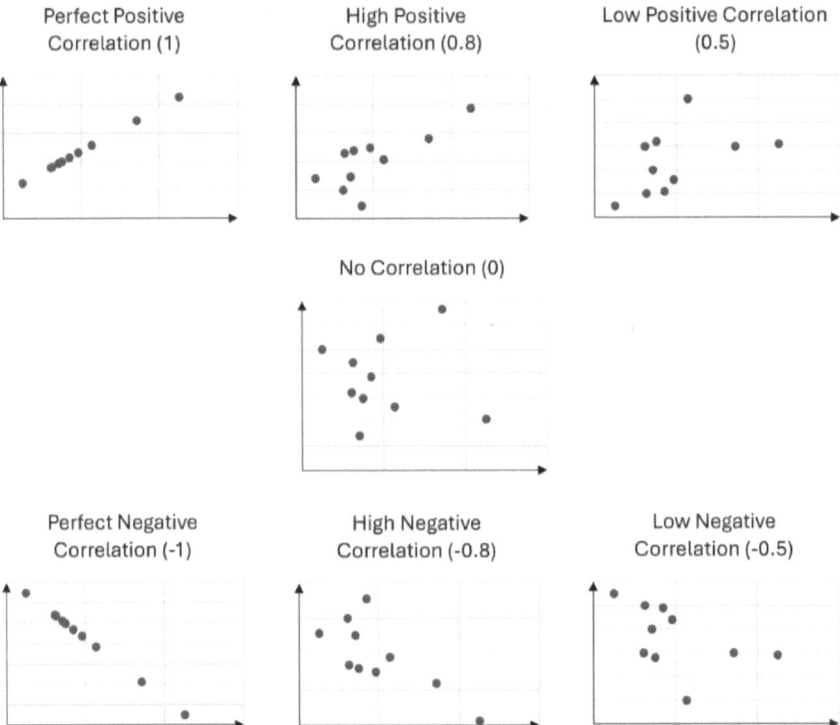

Figure 7.1 Seven distinct scatter plots illustrate different correlation coefficients, ranging from 1 to -1.

positive coefficients of 0.8 and 0.5 (top row) and negative coefficients of -0.8 and -0.5 (bottom row). The center chart shows no discernible trend, with a correlation of 0.

A lack of correlation does not necessarily imply that there is no relationship between the variables; it simply suggests that the connection is either unclear or not statistically significant. A simple example of this is the concept of distance from a central point. Imagine a child walking between two houses, with a stick placed at the midpoint of the path (Figure 7.2). The stick represents the zero point, with movement to the right considered positive and movement to the left considered negative, similar to a number line. One variable is the child's location at a given moment, and the other is her distance from the stick.

There is clearly a relationship between these two variables, but it is neither purely positive nor negative. As she moves to the right, her location (the first variable) increases, and her distance (the second variable) also increases. Conversely, as she moves to the left, her location decreases (since it becomes negative), but her distance still increases. The relationship is positive while she walks to the right and negative while she walks to the left. Despite this clear dependency, the overall correlation between the two variables – location and distance – will be zero, as the positive and negative trends cancel each other out.

Figure 7.2 A diagram depicting two houses with a pole positioned at the midpoint of the path. The midpoint represents the zero point, with movement to the right considered positive and movement to the left considered negative, akin to a number line.

7.1.1 Calculating Pearson's correlation coefficient

We will demonstrate the calculation of Pearson's correlation coefficient using the data analyzed in section 5.2.1, which focused on government stability of the State of Israel. In that chapter, we created two timelines representing the duration and parliamentary support of each Israeli government from 1948 to 2022. Although we initially concluded that it was difficult to assume any correlation between the two phenomena, we left the question open for further investigation. Now, we will revisit this data to explore the potential relationship using Pearson's correlation coefficient.

Pearson's correlation coefficient quantifies the relationship between two variables using the metrics we studied in Chapter 6: **mean** and **standard deviation**. To begin, we first calculate these two measures for each of the phenomena under consideration. By standardizing the data in this way, Pearson's coefficient enables us to compare the degree of association between the two variables, regardless of their units or scales.

Questions

1. Israeli governments:
   a. *Calculate the mean and standard deviation of the duration of Israeli governments (Data available at https://zefsegal.com/computational-literacy-for-the-humanities/).
   b. *Calculate the mean and standard deviation of the maximum number of parliamentary members that supported the coalition (Data available at https://zefsegal.com/computational-literacy-for-the-humanities/).

In the second step of the calculation, we compute the **covariance**, which measures how two variables change together.

- For each data point (in this case, each government), calculate the deviations of X (duration) and Y (members) from their respective means.
- Multiply the deviations for each pair.
- Sum the products of these deviations for all observations.
- Finally, divide the sum by the number of points (governments).

This can be calculated in Google Sheets using the `covariance.p(___)` function. In the parentheses, select one column for X, and another for Y in the parentheses.

Questions

2. *Calculate the covariance of the duration and the parliamentary support of Israeli governments.

Covariance is positive when two variables are positively correlated and negative when they are negatively correlated. Pearson's correlation coefficient is essentially a normalized form of covariance, calculated by dividing the covariance by the product of the standard deviations of both variables: $\dfrac{covariance(X,Y)}{SD(X) \cdot SD(Y)}$. In Google Sheets, this can be calculated using the `correl()` function. In our case study, the corresponding coefficient is $\dfrac{54.16}{13.28 \cdot 13.53} = 0.3$.

Table 7.1 offers a rough interpretation of the correlation between two phenomena. In this case study, the coefficient is 0.3, indicating a weak positive correlation. This suggests that larger coalitions tend to survive slightly longer on average than smaller ones, though the relationship is not strong. This could explain why it was difficult to detect a clear pattern in Chapter 5.

In 1903, Karl Pearson published the findings of a study that analyzed hereditary traits in hundreds of families (Pearson and Lee 1903). He focused on various physical characteristics, such as height and arm length. One key aspect of the study was collecting the height data of 1,078 pairs of fathers and sons (you can access the data at https://zefsegal.com/computational-literacy-for-the-humanities/).

Questions

3. *Calculate the correlation coefficient between the heights of fathers and sons.

The moderate correlation coefficient suggests that height is a hereditary trait, though the relationship is far from perfect. As we know, a son's height is influenced by multiple factors, including the height of the mother. Therefore, it is expected that the correlation between fathers' and sons' heights would not be strongly positive.

7.1.2 *Distant reading: word frequency and length in Wharton's and Stein's texts*

For our final discussion of Pearson's correlation coefficient, we return to the novels and autobiographies of Edith Wharton and Gertrude Stein, which were explored

Table 7.1 Interpreting correlation coefficients

Correlation coefficient	Interpretation
1	Perfect positive correlation
Between 0.5 and 1	Strong positive correlation
Between 0 and 0.5	Weak positive correlation
0	No correlation
Between -0.5 and 0	Weak negative correlation
Between -1 and 0.5	Strong negative correlation
-1	Perfect negative correlation

in Chapter 6. We aim to investigate whether there is a correlation between word length and word frequency in these texts. Previously, we generated frequency lists for each of the four books, listing all word types along with their frequency of occurrence. Now, we will calculate the length of each type and compute the correlation coefficient between word length and frequency.

Questions

4. Word frequency and word length:
    a. Open the frequency lists created in section 6.4.1 in Google Sheets.
    b. Add a new column to calculate the length of each word type, using the formula =LEN() to compute the length for each entry.
    c. Find the correlation coefficient between the values in column A (word frequency) and column C (word length) using the CORREL() function.

The results should be as follows: the correlation coefficient between word frequency and word length in Wharton's Autobiography is -0.1046, in Wharton's Novel is -0.1195, in Stein's Autobiography is -0.1425, and in Stein's Novel is -0.2001. The negative correlation coefficients in all the books are expected: longer words tend to be less frequent, a general linguistic phenomenon not limited to these texts. Additionally, all coefficients fall between -0.5 and 0, indicating a weak correlation. However, there is a noticeable difference between Wharton's and Stein's works – Stein's texts show a stronger correlation, meaning longer words appear even less frequently in her works compared to Wharton's. This aligns with our earlier findings that Stein's vocabulary is more limited and simpler.

As we recall, according to Zipf's law, between 40% and 60% of the words in any text typically appear only once. In Wharton's autobiography, for instance, approximately 54% of the words (6,681) are hapax legomena (words that appear only once). These words generally include rare terms, as well as words whose low frequency is specific to this text. In terms of length, the set of hapax words is diverse, containing both short and long words. Consequently, they may distort the measurement of the correlation between word length and frequency.

The following scatter plot (Figure 7.3) illustrates the relationship between word length and frequency in Wharton's autobiography. As shown on the left end of the plot, the lengths of unique words span the entire Y-axis, from single letters (e.g., the digit '9') to longer words like 'contemplator-and-appreciator' (28 letters). This wide variation may obscure the true correlation between word length and frequency. To get a clearer picture, we should exclude the hapax legomena and recalculate Pearson's correlation.

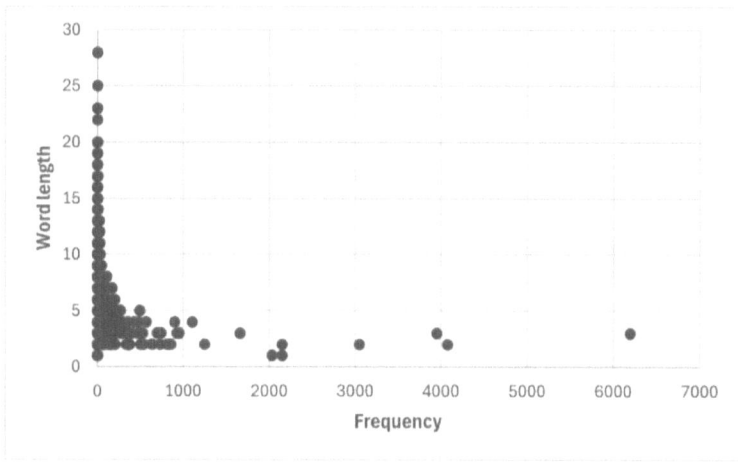

Figure 7.3 A scatter plot depicting word length and frequency in Wharton's autobiography.

We will use the `filter` function to remove the hapaxes. This function requires two arguments: (1) the range of data to filter (e.g., `A1:C50000`), and (2) the filter condition, which applies to one column in the range. In our case, we will filter the data to include only those rows where the value in column A (frequency) is greater than 1, using the condition `A1:A50000>1`.

In cell F1 of the spreadsheet, enter the following filter function: = `filter(A1:C50000, A1:A50000>1)`. This formula filters the data from columns A to C, copying only rows where the frequency (in column A) is greater than 1. The filtered data will appear in columns F through H.

Questions

5. Calculate Pearson's coefficient for the filtered data, for each of the four books.

The resulting Pearson's correlation coefficients for word length and frequency are as follows: Wharton's autobiography shows a coefficient of -0.141, while her novel has a coefficient of -0.1513. For Stein, her autobiography yields a coefficient of -0.1742, and her novel shows the lowest correlation with a coefficient of -0.2303. Although the correlation remains a weak negative one, it is stronger than the previous results, suggesting a clearer inverse relationship between word length and frequency after removing the hapaxes.

Table 7.2 CNN exit polls, 2016, November 23

	Voting for Hillary Clinton	*Voting for Donald Trump*
Married	6,375	7,354
Unmarried	5,537	3,725

Source: https://edition.cnn.com/election/2016/results/exit-polls.

7.2 Chi-Square test for independence: understanding categorical associations

Often, we seek to determine whether there is a statistical dependency between two variables with nominal or qualitative categories. For instance, is there a connection between a movie's genre and the gender of its director? (Günsberg 2004). Did women vote differently from men for the Nazi party in the March 1933 Reichstag elections? (Boak 1989). Can an author's writing style be characterized by the linguistic category of the last word used in their sentences? (Kenny 1982, 111).

Table 7.2 shows the results of a survey conducted by CNN among voters on the 2016 US Election Day:

We might suspect that a voter's marital status influences their choice of presidential candidate. But how can we confirm this intuition? This question is similar to those we explored in the previous section, except that here, our two variables – marital status and candidate choice – are qualitative categories rather than quantitative ones, as seen in the examples of Israeli governments, father–son height pairs, or word length and frequency correlations.

When two variables are completely **independent**, the occurrence of one does not influence the other. For example, if 50% of people are male and 40% prefer coffee, we would expect around 20% (50% × 40%) of the total population to fall into the "Male & Coffee" category by chance. Similarly, the other categories, such as "Male & Tea" or "Female & Coffee", would follow this kind of proportional probability.

Similarly, we can check if the values in Table 7.2 align with the expected results based on the product of the relevant ratios (married/unmarried and Clinton/Trump). However, does this necessarily indicate dependency or independence between the variables? Differences could simply result from a disparity in the number of married versus unmarried voters or from small sample variations, rather than an inherent relationship. The key question is: are the observed differences in our data significant enough to suggest a dependency between the variables in the broader population?

In 1900, Karl Pearson published a seminal paper titled "On the Criterion that a Given System of Deviations from the Probable in the Case of a Correlated System of Variables is Such that it Can Be Reasonably Supposed to Have Arisen from Random Sampling" (Pearson 1900). In this paper, Pearson introduced a test to determine whether differences between results could be attributed to random variation within a sample or are statistically significant. This test became known as the **Chi-Square Test**, named after the Greek letter χ (chi).

Table 7.3 Expected values of the 2016 CNN election polls

	Voting for Hillary Clinton	*Voting for Donald Trump*	*Sum*
Married	6,375	7,354	13,729
Unmarried	5,537	3,725	9,262
Sum	11,912	11,079	22,991

The Chi-Square formula is given by $\chi^2 = \sum_i \dfrac{(O_i - E_i)^2}{E_i}$, where O_i represents the observed frequencies and E_i represents the expected frequencies, assuming the data is distributed according to population proportions. The formula works as a summation loop, where for each observed result, the quotient of $(O_i - E_i)^2$ and E_i is added to the total. The variable i simply represents the iterations of the loop and has no inherent meaning.

Let's apply the method using CNN's 2016 survey results. First, sum the data for each row, column, and the entire Table 7.3.

The percentage of voters for Clinton is $\dfrac{11912}{22991}$ (approximately 51%). Assuming no dependency, the expected number of married voters for Clinton is $\dfrac{11912}{22991} \cdot 13729 = 7113$. The expected frequency in each cell is calculated by multiplying the row and column sums and dividing by the total number of observations.

Accordingly, we obtain the following expected values assuming no dependency between marital status and voting choice:

- The expected number of married voters for Clinton is $\dfrac{11912}{22991} \cdot 13729 = 7113$.

- The expected number of unmarried voters for Clinton is $\dfrac{11912}{22991} \cdot 9262 = 4799$.

- The expected number of married voters for Trump is $\dfrac{11079}{22991} \cdot 13729 = 6616$.

- The expected number of unmarried voters for Trump is $\dfrac{11079}{22991} \cdot 9262 = 4463$.

The expected table can be seen in Table 7.4.

Using the Chi-Square formula:

$$\chi^2 = \sum_i \frac{(O_i - E_i)^2}{E_i} = \frac{(6375 - 7113)^2}{7113} + \frac{(7354 - 6616)^2}{6616}$$
$$+ \frac{(5537 - 4799)^2}{4799} + \frac{(3725 - 4463)^2}{4463} = 395.$$

Table 7.4 Expected values of the 2016 CNN election polls

	Voting for Hillary Clinton	*Voting for Donald Trump*
Married	7,113	6,616
Unmarried	4,799	4,463

Table 7.5 Minimum Chi-Square values for different degrees of freedom (d) and probabilities (0.05, 0.01, 0.001)

d	*0.05*	*0.01*	*0.001*
1	3.841	6.635	10.828
2	5.991	9.21	13.816
3	7.815	11.345	16.266
4	9.488	13.277	18.467
5	11.07	15.086	20.515
6	12.592	16.812	22.458
7	14.067	18.475	24.322
8	15.507	20.09	26.125
9	16.919	21.666	27.877
10	18.307	23.209	29.588

To interpret the value of Chi-Square, we need to calculate the degrees of freedom for our variables, as this helps determine whether the results could be due to chance. Consider a group of 15 people sitting in a row: the first 14 can choose their seats, but the 15th person must sit in the remaining seat, giving a degree of freedom of 14. Similarly, when fitting 15 numbers to a given average, we can freely select 14, but the 15th is determined by the average. Similarly, the degree of freedom for a table is given by the formula: $(number\ of\ rows - 1) \cdot (number\ of\ columns - 1)$. In the 2016 election polls, the degrees of freedom is (2-1)x(2-1)=1.

Pearson demonstrated in his 1900 paper that by combining the Chi-Square value and the degrees of freedom, we can calculate the probability that the observed data were produced by chance – meaning, without any dependency between the variables. Table 7.5, originally discovered by Pearson, lists the minimum Chi-Square values required for different degrees of freedom and various probabilities. If the value we obtain exceeds the value in the table, we can reasonably conclude that there is a significant relationship between the two variables.

In the previous example, we obtained a Chi-Square value of 395 with 1 degree of freedom. This is significantly higher than any values listed for 1 degree of freedom (second row), indicating with over 99.9% confidence that there is a non-random connection between marital status and voting patterns in CNN's survey. The probability of the results being random can be calculated directly in Google Sheets using the chisq.test() function, with the observed and expected tables as arguments. In this case, the probability is $9.03 \cdot 10^{-88}$, an extremely low value.

Table 7.6 Distribution of first 100 words according to parts of speech

	Noun	*Verb*	*Other*	*Total*
Rhetoric	28	32	40	100
R. to Alexander	27	52	21	100
Total	55	84	61	200

The Chi-Square test can be applied to any number of categories and groups. In the previous example, we used a 2x2 table (two marital statuses and two voting choices), but more complex tables can include additional rows and columns. However, this test has its limitations:

- The Chi-Square test should only be used with whole numbers, not percentages or ratios.
- Categories must be mutually exclusive, with no overlap between cells.
- The test becomes unreliable when any expected value is less than 5.
- While the test identifies relationships, it does not explain the cause of these relationships, similar to other calculations discussed in this book.

7.2.1 Case study: Author identification through parts of speech analysis

The Greek philosopher Aristotle (384–322 BCE) is traditionally attributed with two works on rhetoric: *Rhetoric* (Rhētorikḗ) and *Rhetoric to Alexander* (Rhetorica ad Alexandrum). However, modern scholars believe *Rhetoric to Alexander* was actually authored by Anaximenes of Lampsacus (380–320 BCE). This conclusion was reached by analyzing the writing styles of both texts, revealing significant differences that suggest different authorship (Kenny 1982, 111).

Table 7.6 compares the first one hundred sentences from each work. Each sentence is classified based on the part of speech of its final word: sentences ending in a noun, sentences ending in a verb, and sentences ending in other types of words.

Questions

6. Authors and parts-of-speech:
   a. *Calculate the Chi-Square value.
   b. *What are the degrees of freedom?
   c. *Calculate the probability that the difference between the two texts occurred by chance.

Typically, a Chi-Square test requires a significance level of less than 5% (0.05) to confidently claim that the difference is non-random. In the previous example, the probability obtained was far below this threshold, allowing us to assert with certainty that the texts reflect different writing styles, strongly suggesting different authors.

7.2.2 Part-of-speech tagging with Python

A key step in analyzing the distribution of words by their lexical category is **part-of-speech (POS) tagging,** a process in which each word is labeled with a **tag** that reflects its lexical category. While it's possible to tag manually, using software is much more practical for longer texts. We will use the `pos_tag` function from the `nltk` library for automatic tagging. The process starts by choosing a set of tags, with the Penn Treebank set being the most common one, comprising 36 tags (www.ling.upenn.edu/courses/Fall_2003/ling001/penn_t reebank_pos.html).

In this section, we will primarily focus on nouns and verbs:

- **NN**: Noun, singular or mass (e.g., 'cat')
- **NNS**: Noun, plural (e.g., 'cats')
- **NNP**: Proper noun, singular (e.g., 'James')
- **NNPS**: Proper noun, plural (e.g., 'Vikings')
- **PRP**: Personal pronoun (e.g., 'she')
- **PRP$**: Possessive pronoun (e.g., 'his')
- **VB**: Verb, base form (e.g., 'get')
- **VBD**: Verb, past tense (e.g., 'got')
- **VBG**: Verb, gerund or present participle (e.g., 'getting')
- **VBN**: Verb, past participle (e.g., 'gotten')
- **VBP**: Verb, non-3rd person singular present (e.g., 'get')
- **VBZ**: Verb, 3rd person singular present (e.g., 'gets')

The tagging system includes finer distinctions beyond basic parts of speech. For example, verbs are categorized into six subtypes based on tense and form, while nouns are classified into singular, plural, common, and proper nouns. Despite these divisions, tags within the same lexical category (e.g., verbs) share similar prefixes, such as "VB". For our current case study, we will treat all verb subtypes as "verbs" and all noun subtypes (including pronouns) as "nouns". For more on tagging, you can explore the NLTK guide (www.nltk.org/book/ch05.html).

To use natural language processing tools from the `nltk` library, we first import the library and then download the `punkt` module for tokenization and the `averaged_perceptron_tagger` for tagging words based on their parts of speech. This step only needs to be performed once at the beginning of the code.

```
import nltk
nltk.download("punkt")
nltk.download("averaged_perceptron_tagger")
```

Next, we execute two processes: tokenization and POS tagging. In the following block, we begin with a string variable `testext`. The string is passed as an argument to the `word_tokenize` function, which returns a list of words (`testextwords`). Note that tokenization separates words from any attached punctuation.

```
testext = "You cannot shake hands with a clenched fist."
testextwords = nltk.word_tokenize(testext)
print(testextwords)
```

['You', 'can', 'not', 'shake', 'hands', 'with', 'a', 'clenched', 'fist', '.']

The result is then passed as an argument to the `pos_tag` function, which tags each word by its part of speech and returns a list of tuples, where the first element is the word, and the second is its part-of-speech tag.

```
tagged = nltk.pos_tag(testextwords)
print(tagged)
```

```
[('You', 'PRP'), ('can', 'MD'), ('not', 'RB'),
('shake', 'VB'), ('hands', 'NNS'), ('with', 'IN'),
('a', 'DT'), ('clenched', 'JJ'), ('fist', 'NN'),
('.', '.')]
```

Experimenting with this process can help before proceeding with the main exercise.

To access an element in the list, use an index. Remember that indexing starts at 0. For instance, to access the first word–tag pair in the list, use [0]. To access the first part of the tuple or the second part, use [0][0] or [0][1], respectively.

```
print(tagged[0])
print(tagged[0][0])
print(tagged[0][1])
```

```
('You', 'PRP')
You
PRP
```

A loop can be used to iterate over the list of tuples in two ways: either over the tuples themselves or over their elements (see also section 5.3.2). Note the difference between the two loops below:

```
for x in tagged:
    print(x)
for i,j in tagged:
    print(i,j)
```

Now that our text is tagged with parts of speech, we can count the number of nouns, verbs, and other words. To do this, we'll use a loop to iterate over the list of tagged tuples. Before beginning, we define variables to serve as counters for each category and initialize them to zero:

```
nouns = 0
verbs = 0
other = 0
```

In the `for` loop, we use two variables: `word` for the first element in the tuple and `tag` for the second, which we will use for counting. We distinguish between four cases:

- The tag starts with "N" (indicating a noun).
- The tag is "PRP" (personal pronouns like 'he', 'she', 'they').
- The tag starts with "V" (indicating a verb).
- All other cases.

Since we do not differentiate between nouns and pronouns, the first condition will include both, separated by the `or` operator. To access the first letter of the tag, we use indexing.

```
for word, tag in tagged:
    if tag[0] == "N" or tag == "PRP":
        nouns += 1
```

```
elif tag[0] == "V":
    verbs += 1
else:
    other += 1
```

Note that in this code, instead of writing `nouns = nouns + 1` we use a short-hand method to increment the counters by 1: `nouns += 1`.

At the end of the loop, the output is 3 variables, reflecting the number of nouns, verbs, and other parts of speech in the text.

7.2.3 Distant reading: distribution of parts of speech in Wharton and Stein

So far, we have identified many differences in the writing styles of Wharton and Stein's novels and autobiographies:

Average word frequency: Wharton's vocabulary is more diverse than Stein's, which is characterized by a more limited vocabulary and frequent repetition.

Average sentence length: The sentences in both novels as well as Stein's auto-biography are similar in length. The sentences in Wharton's autobiography are much longer.

Average word length: The average word length in all books is similar. The standard deviation of word length in Wharton's books is higher, indicating a wider range of lengths. Additionally, the correlation coefficient showed that long words tend to appear slightly more frequently in Wharton's works.

In this section, we will use Python to analyze another stylistic feature: the distribution of words by parts of speech. This approach will help us delve deeper into the linguistic patterns in Wharton's and Stein's works, categorizing words into nouns, verbs, adjectives, and other parts of speech. By comparing the usage of these categories, we can uncover further insights into the distinct writing styles of the two authors.

Questions

7. For each of the four books:
   a. Read the content of the book into the variable rawtext.
   b. *Tokenize rawtext into the variable tokens and use it as an argument for pos_tag. Store the POS-tagged text in the variable tagged.
   c. *Calculate the 3 values (nouns, verbs, and other parts of speech) for each text.
   d. Save all the results to a text file named "POS-counts.txt".

At the end of the process, we will obtain a text file containing the frequency of nouns, verbs, and other words for each of the books. We will use this output to perform three comparisons:

- Wharton's autobiography vs. Stein's autobiography
- Wharton's novel vs. Stein's novel
- Wharton's works vs. Stein's works

For each comparison, we will define a table similar to the one created for *Rhetoric* and *Rhetoric to Alexander*.

Questions

8. Nouns and verbs:
   a. *Calculate the ratio of nouns to verbs for each text.
   b. *Perform the Chi-Square test for each comparison (the probability suffices).

The Chi-Square results show that the differences in Wharton and Stein's writing styles, as reflected by their use of parts of speech, are significant. The probability that these differences are random is close to zero. The noun-to-verb ratio reveals that Wharton tends to use far more nouns than Stein, further emphasizing the richer and more varied language in Wharton's works. Additionally, both authors use relatively more nouns in their autobiographies.

7.3 Spearman's correlation: assessing relationships between ranked variables

Pearson's correlation coefficient and the Chi-Square test help us assess the dependency between two variables based on their absolute values. However, in many cases, the order of data points is more significant than the actual values themselves. For example, in a sports event, the finishing order is more important than each athlete's score. Similarly, economic rankings, the sequence of chapters in a novel, or a royal lineage are best understood through ranked information rather than numerical data.

We will use the historical ranking of municipalities in Germany as an illustrative example (see Table 7.7).

The ranked lists from 1880, 1910, and 1939 show significant changes over time. While Berlin remains the largest city throughout, other cities like Wrocław drop from third place in 1880 to seventh in 1910 and ninth in 1939. Meanwhile, Essen, absent from the 1880 list, rises to 13th place in 1910 and seventh by 1939. It's important to note that we are only considering rankings here, ignoring population sizes. For instance, Berlin had 1.1 million residents in 1880 and nearly 4.5 million in 1939.

Table 7.7 The 20 largest municipalities in Germany in 1880, 1910, and 1939

Rank	1880	1910	1939
1	Berlin	Berlin	Berlin
2	Hamburg	Hamburg	Vienna
3	Wrocław	Munich	Hamburg
4	Munich	Leipzig	Munich
5	Dresden	Dresden	Cologne
6	Leipzig	Cologne	Leipzig
7	Cologne	Wrocław	Essen
8	Königsberg	Frankfurt am Main	Dresden
9	Frankfurt am Main	Düsseldorf	Wrocław
10	Hanover	Nuremberg	Frankfurt am Main
11	Stuttgart	Charlottenburg	Dortmund
12	Bremen	Hanover	Düsseldorf
13	Danzig	Essen	Hanover
14	Strasbourg	Chemnitz	Stuttgart
15	Nuremberg	Stuttgart	Duisburg
16	Magdeburg	Magdeburg	Nuremberg
17	Bremen	Bremen	Wuppertal
18	Düsseldorf	Königsberg	Königsberg
19	Chemnitz	Rixdorf	Bremen
20	Elberfeld	Stettin	Chemnitz

Sources: Kaiserliches (1880–1918); Statistisches (1919–1941/2).

Instead of analyzing the cities' growth rates, we are focusing on changes in their status and the correlation between different rankings. The assumption is that only significant shifts in the urban landscape would lead to the decline of large cities and the rise of smaller ones. By examining these shifts, we aim to understand how various factors may have influenced the relative rankings of municipalities over time.

7.3.1 Calculating Spearman's correlation coefficient

In 1904, English psychologist Charles Edward Spearman (1863–1945) introduced **Spearman's rank correlation coefficient** in his paper "General Intelligence, Objectively Determined and Measured" (Spearman 1904). This measure evaluates the correlation between ranked lists of two variables. Unlike Pearson's correlation, which examines explicit values, Spearman's coefficient focuses on the relationship between ranks. It is especially useful when the relative order of data points, rather than their specific values, is the key to understanding the relationship between two variables.

The formula for this coefficient is given by $r = 1 - \dfrac{6 \sum (d^2)}{n^3 - n}$, where d represents the difference in rankings, and n is the number of observations. The squares of the differences are summed, and this sum is used in the formula. The steps to calculate this are as follows:

I.   Preparation: You have two sets of data. Assign ranks to each data point in both lists. The lowest value receives a rank of 1, the next lowest a rank of 2, and so on. For tied values, assign the average rank.
II. Calculate the difference in ranks (d): Subtract the rank of one variable from the other for each pair of data points.
III. Square the differences (d^2): Square each difference to remove negative values.
IV. Sum the squared differences (Σd^2): Add all the squared differences.
V.   Apply Spearman's rank formula: Use the formula to compute the coefficient.

Similarly to variance, Pearsons's correlation coefficient, and the Chi-Square test, Spearman's Rank Correlation Coefficient measures relative dispersion by calculating distances. When measuring distances by subtracting values, both positive and negative differences can arise, which could cancel each other out. To prevent this, all four metrics square the differences, ensuring that both positive and negative deviations contribute equally, thereby providing a more accurate measure of dispersion.

Like Pearson's correlation coefficient, Spearman's correlation coefficient ranges between -1 and +1. A value of 0 indicates no correlation between the two rankings. A result of +1 signifies a perfect positive correlation, meaning the rankings are identical. Conversely, a value of -1 represents a perfect negative correlation, where the rankings are exactly reversed.

Spearman's rank correlation coefficient is a valuable tool, but it has some limitations:

- **Rankable Data**: The test requires data that can be ranked, making it unsuitable for categorical data without inherent order.
- **Sample Size**: A sufficient sample size is needed for reliability; small samples (fewer than 10) may yield unreliable results.
- **Handling Ties**: The test is less precise when there are many tied ranks, as it assumes unique rankings.
- **List Compatibility**: Spearman's correlation requires matching data points in both lists. If most points overlap, unpaired items can be excluded.

7.3.2 Case study: Urban landscape in turn-of-the-century Germany

We will use Spearman's correlation to compare each consecutive pair of ranked lists from Table 7.7 (1880 to 1910, and 1910 to 1939). If no significant changes occurred in either period, the correlation coefficients should be nearly identical and close to 1. However, dramatic transformations such as industrialization and imperialism in the 1880s and 1890s, and the impact of World War I in the 1910s altered the German urban landscape. This statistical analysis will reveal which events had a more decisive influence on the shifts in urban space.

Table 7.7 includes three ranked lists, but several cities do not overlap between the lists. As noted earlier, Spearman's correlation coefficient is not effective for values that are not shared by both lists. Therefore, we will filter out the cities that do not appear in both rankings and calculate the differences in rank for the cities that remain (Table 7.8).

According to Table 7.8, the sum of squared differences is 162 for the 1880–1910 comparison and 74 for the 1910–1939 comparison. With 16 items in both cases, we apply Spearman's rank correlation coefficient. For the 1880–1910 comparison, $r = 1 - \dfrac{6 \times 162}{16^3 - 16} = 0.76$ and $r = 1 - \dfrac{6 \times 74}{16^3 - 16} = 0.89$.

Both coefficients reflect a relatively high level of similarity between consecutive rankings, yet the difference between the two is quite significant. This suggests that the changes in urban patterns following Germany's unification in 1871, along with industrialization and imperialism, had a greater impact on urban development than the effects of World War I, the Weimar Republic's instability, or the economic crisis of 1929, which occurred between 1910 and 1939.

While correlation coefficients provide a high-level view of ranking shifts, they may not reveal outliers or specific trends. A deeper analysis of the lists can shed more light on the changes taking place. For instance, between 1880 and 1910, the notable decline of eastern cities like Wrocław and Königsberg contrasts with the rise of western cities such as Düsseldorf and Cologne, as well as the ascent of commercial and industrial centers like Nuremberg, Leipzig, and Chemnitz. These shifts illustrate more granular changes in urban dynamics during this period.

Table 7.8 A filtered list of urban rankings comparing 1880 to 1910 and 1910 to 1939

	1880	*1910*	*d2*		*1910*	*1939*	*d2*
Berlin	1	1	0	Berlin	1	1	0
Hamburg	2	2	0	Hamburg	2	2	0
Wrocław	3	7	16	Munich	3	3	0
Munich	4	3	1	Leipzig	4	5	1
Dresden	5	5	0	Dresden	5	7	4
Leipzig	6	4	4	Cologne	6	4	4
Cologne	7	6	1	Wrocław	7	8	1
Königsberg	8	16	64	Frankfurt am Main	8	9	1
Frankfurt am Main	9	8	1	Düsseldorf	9	10	1
Hanover	10	11	1	Nuremberg	10	13	9
Stuttgart	11	13	4	Hanover	11	11	0
Bremen	12	15	9	Essen	12	6	36
Nuremberg	13	10	9	Chemnitz	13	16	9
Magdeburg	14	14	0	Stuttgart	14	12	4
Düsseldorf	15	9	36	Bremen	15	15	0
Chemnitz	16	12	16	Königsberg	16	14	4

7.3.3 Case study: Pierre Corneille and historical change in writing style

The early 17th century in French literature was marked by significant, conscious changes in the language of the educated classes, as noted by Frederick Morris Warren (1890, 194). Pierre Corneille (1606–1684), one of the greatest French dramatists of the time, was particularly sensitive to these evolving linguistic trends, due to his dependence on popular favour. Warren's analysis of Corneille's plays highlights changes in form, tense, and gender across his works, supporting his claim that "at times of linguistic and grammatical change the statistical method can be safely followed" (Warren 1890, 198). In this exercise, we will apply Spearman's correlation coefficient to examine changes in the average number of words per verse in ten of Corneille's plays, as listed in Table 7.9.

Questions

9. Plays, years, and words:
   a. Rank the plays by year of composition and by the average number of words per verse.
   b. *Calculate Spearman's rank correlation coefficient for the two ranking lists.

Much like Warren's linguistic analysis, Spearman's rank correlation coefficient offers insight into Corneille's evolving style. The coefficient reveals a strong positive correlation between the year of composition and the number of words per verse, suggesting that as Corneille aged, his verses contained more words. Given that he wrote in French Alexandrine verse – where each line of verse consists of exactly 12 syllables – this increase in word count implies that the average word

Table 7.9 A selection of Corneille's plays, showing the year of release and average words per verse

Play	Year	Words per verse
Melite	1629	8.93
La Galerie du Palais	1632	9.02
VIllusion Comique	1635	9.15
Horace	1640	9.26
Rodogune	1644	9.15
Andromede	1650	9.2
Sertorius	1662	9.22
Agesilas	1666	9.32
Pulcherie	1672	9.48
Surena	1674	9.53

Sources: Müller (1967); Kenny (1982, 74).

length shortened. Thus, Corneille's later plays featured more words per sentence, but the words themselves were shorter, reflecting a shift in his writing style over time.

7.4 Linear regression: modeling relationships between variables

Correlation is a symmetric relationship: if height is correlated with weight, then weight is equally correlated with height, and the strength of the correlation is the same in both directions. However, in some cases, we are more interested in asymmetric relationships between two phenomena. For example, in Pearson's foundational research, we might ask how a son's height changes based on his father's height. Similarly, regarding Israel's political stability, we might ask how much tenure increases with each change in coalition size.

Of course, the ability to estimate one variable using another relies on a fairly strong correlation, whether positive or negative. We will demonstrate this with two clearly dependent variables: the age at which British monarchs ascended to the throne and the length of their reign (data available at https://zefsegal.com/comput ational-literacy-for-the-humanities/). Such dependency doesn't always exist in monarchic regimes. For example, in the Roman Empire, emperors had short and unstable reigns, often lasting only a few years. In fact, one-third of the emperors lasted less than one year. However, since the Act of Settlement (1701) and the Act of Union (1707), British monarchs have experienced more regular successions, primarily due to natural causes, with Edward VIII's abdication being a notable exception.

The scatter plot shows a strong negative correlation: the older a monarch was at accession, the shorter their reign. Assuming a nearly linear relationship, we could estimate the length of reign based on the monarch's age at accession. For instance, a monarch crowned at age 50 might be expected to reign for about 20 years. While predicting hypothetical reigns has limited practical use, this method illustrates how historians can "fill" gaps in historical data using statistical predictions when strong correlations between variables exist.

Returning to the example of the British royal family, how do we determine the location and direction of the imaginary line based on the scatter plot? The most common method is **linear regression**, a technique developed by Karl Pearson. Unlike other statistical measures, linear regression produces an equation that describes the relationship between two variables. This line, as seen in Figure 7.4, can be visualized and calculated in Google Sheets using **trendlines**, as explained in section 5.2.1.

The goal of any regression is to find an equation that best represents the relationship between pairs of values in a scatter plot. However, most data do not perfectly fit a straight line or curve, as seen in Figure 7.4, where only few points lie directly on the line. This brings up the question of what constitutes the "best fit" and how to calculate the deviation between the line or curve and the actual data points.

The accepted method for determining the "best fit" line is the **method of least squares**, originally developed in astronomy and geodesy to compare mathematical

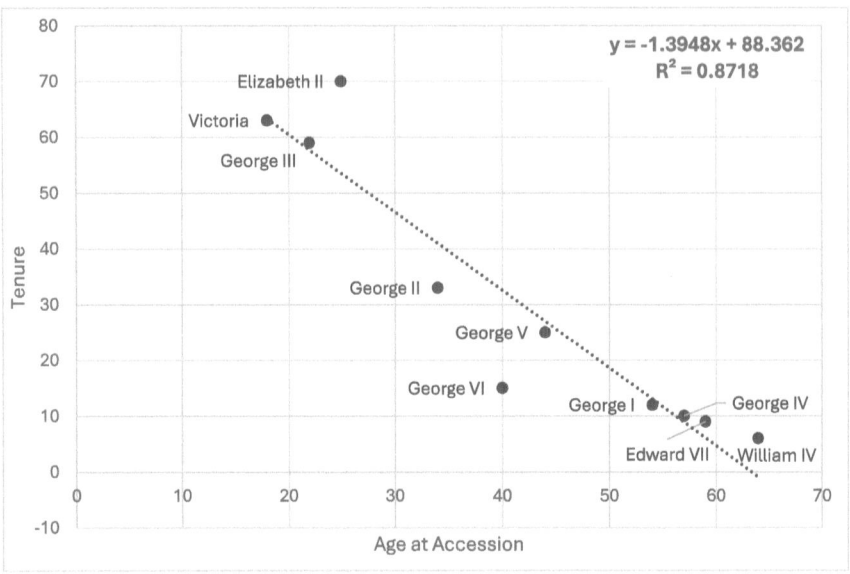

Figure 7.4 A scatter plot illustrating the age at which British monarchs (since 1714) ascended to the throne and the length of their reign.

models for identical data sets, such as celestial positions. Two mathematicians, Adrien Marie Legendre (1752–1833) and Carl Friedrich Gauss (1777–1855), both claimed credit for its invention. Legendre first published the formula in 1805 in a paper on comets, while Gauss included it in his 1809 paper on the dwarf planet Ceres. Today, the method is commonly attributed to Gauss.

Assume we have a scatter plot, like the one linking the length of reign to the age of accession of British monarchs, with an arbitrary line drawn through it (Figure 7.4). Gauss proposed measuring the distance from each data point to this line. The ideal line would have all distances equal to zero, but such a line rarely exists. Instead, Gauss's method seeks to find the line where the sum of the squared distances from each point is minimized, hence the name least squares method.

The equation in the top right corner of Figure 7.4, $y = -1.3948x + 88.362$, represents the "best fit" line calculated using the least squares method. From this equation, we learn that for every additional year in the monarch's age at accession, the length of reign decreases by approximately 1.4 years. Additionally, this formula can be used to predict future reigns. For instance, if a monarch ascended the throne at age 25, like Queen Elizabeth II, their predicted reign would be $-1.3948 \times 25 + 88.362 = 53.492$.

However, Queen Elizabeth II reigned for 70 years, far exceeding the predicted 53.5 years from the regression equation. By observing her distance from the trendline in the plot, we could have identified her as an outlier. Similarly, George

VI, who ascended the throne at age 40, was predicted to reign for 32.57 years but only reigned for 15 years, making him another outlier. Aside from these two, and possibly George II, most monarchs align well with the linear approximation.

The approach to handling outliers varies depending on research goals. In some cases, outliers are removed and analyzed separately. In this example, Queen Elizabeth II's reign can be considered an outlier due to her exceptional longevity, while George VI's early death was influenced by health issues related to smoking. By treating these cases separately, we gain more insight into the individual factors that contributed to their deviation from the overall trend.

However, this approach is limited, as isolating every outlier isn't always feasible. If there are too many outliers, the regression line becomes ineffective. The **Coefficient of Determination** (R^2) helps measure how well one variable predicts another. Expressed as a decimal between 0 and 1, it indicates the percentage of variation in the dependent variable (y) explained by the independent variable (x). If R^2 equals 1, it means one variable perfectly explains the other – every data point fits exactly on the line. If R^2 equals 0, the independent variable doesn't explain any of the variation.

For example, in our case, R^2=0.8718, meaning that 87% of the variation in the length of a monarch's reign can be explained by their age at accession. This tells us that age is a strong predictor of how long a British monarch will reign, but 13% of the variation is due to other factors, such as individual circumstances like Queen Elizabeth II's long life or George VI's health issues.

7.4.1 Distant reading: verb-to-word ratio and literary styles

The relationship between the length of sentences and the number of verbs they contain is an additional measure with which we can gain valuable insights into the linguistic and stylistic dimensions of a text. Verbs, especially lexical ones, signify actions and events. A higher verb-to-word ratio often indicates denser, action-oriented sentences, while a lower ratio suggests more descriptive or reflective prose.

Taking Edith Wharton's autobiography as an example, we will use linear regression to investigate the relationship between the two variables – sentence length and number of verbs – and represent this information in a scatter plot. The scatter plot and the regression line will serve as means of identifying sentences that are typical to the text as well as outliers.

As a first step, we will use Python to calculate the number of words and the number of verbs for each sentence and to create a text file with the results. We will then import the data to Google Sheets, where we will create the scatter plot and calculate the trendline, equation, and R^2 value.

The algorithm of the following program is structured as a loop within a loop. The outer loop iterates through all the sentences in the text, and the inner one iterates through the (word, POS) tuples of each sentence, after it has been POS-tagged. It assumes that the text was read and cleaned and assigned to the variable `cleantext`, that the `nltk` library has been imported, and that `output_file` is a handle to a text file that is open for writing.

```
# Tokenize the text into sentences
sentences = nltk.sent_tokenize(cleantext)

# external Loop: interates through sentences in
the text
for sent in sentences:
    # create lists of words, tokens, (word, POS) tuples
    words_l = sent.split()
    tokens = nltk.word_tokenize(sent)
    tagged = nltk.pos_tag(tokens)
    verbs = 0

# internal Loop: iterates through (word, POS) tuples
in a tagged sentence
    for word, tag in tagged:
        if tag[0] == "V":
            verbs += 1
    if verbs!= 0:
        output_file.write(str(verbs) + "," +
        str(len(words_l)) + "\n")
```

The analysis outcome is depicted in Figure 7.5. The scatter plot illustrates the relationship between the number of verbs (X-axis) and sentence length (Y-axis). The regression line and its equation show that, on average, adding a verb increases sentence length by about five words. This means that verbs comprise roughly one-fifth of the words in Wharton's sentences, regardless of overall sentence length. The R^2 value of 0.628 indicates a reasonably strong relationship between sentence length and the number of verbs.

If we were to continue with a distant reading of the four books, we would perform the same process for the rest of the books and then compare the outcomes. We would find that the verb-to-word ratio in Wharton's writing is lower than that of Stein's and that the coefficient of Wharton's autobiography (4.79) is outstanding relative to the coefficients of the other three books.

However, computational outputs, like the scatter plots or the linear equations, reveal general stylistic patterns but do not capture the nuances of specific phenomena. These measures can be used as indicators of phenomena that warrant further investigation, or, in other words, a **close reading**. In this particular case, we can use them to identify sentences that are typical to the text as well as outliers.

As we discovered in the distant reading of the novel which we conducted in section 6.4, the sentences in Wharton's autobiography are relatively long, with an average length of nearly 33 words and a 30 words median. Moreover, the linear regression calculation revealed that regardless of length, typically

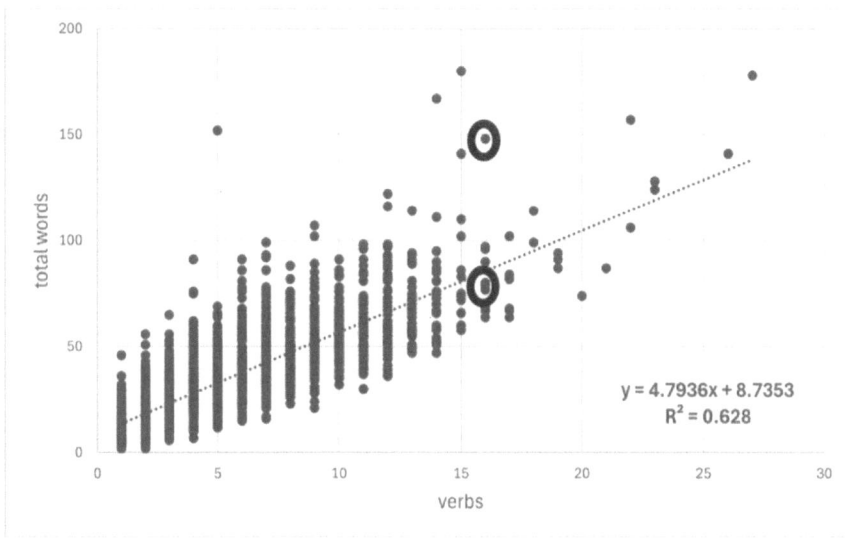

Figure 7.5 Verb-to-word ratio in Edith Wharton's autobiography.

sentences in this book contain aapproximately one verb for every five words. Finally, from the scatter plot in Figure 7.5 we can see that deviations from this trend occur mostly above the trendline, with sentences with a much higher verb-to-word ratio.

The difference between typical sentences and outliers is illustrated by the following examples sentences, corresponding to the two circled data points in Figure 7.5. Each sentence contains 15 verbs (highlighted), yet they vary greatly in length, with the first containing 73 words and the second 141.

(1) I was made to feel afterward that Jusserand and I had failed in our duty in not organizing the party in such a way that each guest should have a few minutes' talk with the great man; for it was inconceivable to my amiable but highly disciplined guests that either the President or his hostess should unintentionally omit a single move of the traditional game they had been invited to play with him.

(2) I am now enumerating only our daily fare, that from which even my tender years did not exclude me; but when my parents "gave a dinner", and terrapin and canvas-back ducks, or (in their season) broiled Spanish mackerel, soft-shelled crabs with a mayonnaise of celery, and peach-fed Virginia hams cooked in champagne (I am no doubt confusing all the seasons in this allegoric evocation of their riches), lima-beans in cream, corn souffles and salads of oyster-crabs, poured in varied succulence from Mary Johnson's lifted cornucopia – ah, then, the gourmet of that long-lost day, when cream

was cream and butter butter and coffee coffee, and meat fresh every day, and game hung just for the proper number of hours, might lean back in his chair and murmur "Fate cannot harm me" over his cup of Moka and his glass of authentic Chartreuse.

Although the two sentences are by no means short and concise, as is typical of Wharton's writing, the difference in verb-to-word ratio is manifested in the amount of detail, as well as the pace. The first sentence is dynamic and event-driven ('fail', 'organize', 'invite', 'play'). Conversely, the goal of the second sentence is to evoke a culinary experience. Here, Wharton provides extensive information about the food served, with the verbs mostly used to support the descriptions ('broiled', 'cooked', 'poured').

However, a close reading also reveals a limitation in the method we adopted. Although the number of verbs is similar across both sentences, treating all verbs equally overlooks crucial distinctions between them. The most significant differentiation lies between lexical verbs – such as 'fail', 'lean', and 'invite' – and auxiliary verbs – like 'had' and 'did' – which often appear alongside lexical verbs ('had failed', 'had been invited', 'did not exclude') and add temporal information. The two types of verbs do not contribute equally to the meaning and complexity of a sentence. Therefore, it is preferable to treat a combination of a lexical verb and auxiliary verb(s) as a single unit, as this generally describes one event. Indeed, in our case, with only 7 lexical verbs in the first example, and 11 in the second, the difference in ratios between the two is reduced, yet the ratio of the second example remains higher.

Despite the critique of our verb-counting method, we have seen that distant reading allows us to uncover general patterns within the text, as well as exceptional cases, guiding us towards a closer reading. By combining both methodologies, we can gain new insights into written texts and their authors' writing styles.

7.5 Correlation does not provide interpretation

It is important to conclude this chapter with a caution regarding quantitative analysis – numbers alone do not provide interpretation, nor can they prove causality. To illustrate this, consider one of the many absurd examples collected by Tyler Vigen on his website, tylervigen.com. These examples highlight spurious correlations – where two unrelated variables seem statistically linked purely by coincidence. This serves as a reminder that correlation alone is not enough to draw conclusions about real-world relationships.

Figure 7.6 shows an almost perfect correlation (0.992558) between the divorce rate in Maine and the per capita consumption of margarine in the USA between 2000 and 2009. Despite the high correlation, it's clear that there is no logical connection between these two variables – just coincidence. This example illustrates that while

Figure 7.6 A chart showing the correlation between divorce rate in Maine and per capita consumption of margarine between 2000 and 2009.

numbers can show relationships, they cannot replace common sense and a proper understanding of the data. **Correlation does not imply causation.**

Glossary

Chi-Square Test A statistical test used to determine if there is a significant association between two categorical variables. It compares observed and expected frequencies in a contingency table.

Coefficient of Determination (R^2) A measure that indicates the proportion of the variance in the dependent variable that is predictable from the independent variable, with values ranging from 0 to 1.

Correlation A statistical measure that describes the strength and direction of the relationship between two variables. Correlation coefficients range from -1 (perfect negative correlation) to +1 (perfect positive correlation), with 0 indicating no correlation.

Covariance A measure of how much two random variables change together. A positive covariance indicates that the variables tend to increase together, while a negative covariance indicates that one increases as the other decreases.

Linear Regression A method for modeling the relationship between two quantitative variables by fitting a straight line (called a regression line) through the data points in a scatter plot.

Part-of-speech (POS) tagging the process of labeling each word in a text with its grammatical category, like noun, verb, or adjective.

Pearson's Correlation Coefficient A statistical measure developed by Karl Pearson that quantifies the linear relationship between two variables, with values ranging from -1 to 1.

Spearman's Rank Correlation Coefficient A non-parametric measure of the strength and direction of the association between two ranked variables.

References

Boak, Helen L. 1989. "'Our Last Hope'; Women's Votes for Hitler: A Reappraisal". *German Studies Review* 12(2): 289–310. https://doi.org/10.2307/1430096

Günsberg, Maggie. 2004. *Italian Cinema: Gender and Genre*. Palgrave Macmillan.

Kaiserliches Statistisches Amt (Eds.). 1880–1918. *Statistisches Jahrbuch für das Deutsche Reich*. Puttkammer & Mühlbrecht.

Kenny, Anthony. 1982. *The Computation of Style: An Introduction to Statistics for Students of Literature and Humanities*. Pergamon Press.

Müller, Charles. 1967. *Étude De Statistique Lexicale: Le Vocabulaire Du Théâtre De Pierre Corneille*. Larousse.

Pearson, Karl, and Alice Lee. 1903. "On the Laws of Inheritance in Man". *Biometrika* 2:357–462. https://doi.org/10.2307/2331507

Pearson, Karl. 1900. "On the Criterion that a Given System of Deviations from the Probable in the Case of a Correlated System of Variables is such that it can be Reasonably supposed to have arisen from Random Sampling". *The London, Edinburgh, and Dublin Philosophical Magazine and Journal of Science* 50(302): 157–175. https://doi.org/10.1080/1478644000 9463897

Spearman, Charles Edward. 1904. "General Intelligence, Objectively Determined and Measured". *American Journal of Psychology* 15: 201–293. https://doi.org/10.2307/1412107

Statistisches Reichsamt (Eds.). 1919–1941/2. *Statistisches Jahrbuch für das Deutsche Reich*. Publisher for Politics and Economics & Hobbing.

Warren, F.M. 1890. "Style and Chronology in Corneille". *The American Journal of Philology* 11(2): 193–199. https://doi.org/10.2307/288013

Solutions to selected questions

1. Israeli governments:
   a. The mean equals 21.58 months, and the standard deviation equals 13.28 months.
   b. The mean equals 77.25 members, and the standard deviation equals 13.53 members.
2. The covariance is 54.16.
3. The correlation coefficient is 0.5.
6. Authors and parts-of-speech:
   a. The Chi-Square value is 10.69.
   b. There are 2 degrees of freedom.
   c. The probability is 0.004753.

7. Distant reading (1):
   b. Tokenization:

```
tokens = nltk.word_tokenize(rawtext)
tagged = nltk.pos_tag(tokens)
```

   c. Counting nouns, verbs, and others:
      i. Wharton's autobiography: 31,517 nouns, 16,119 verbs, 72,130 others
      ii. Wharton's novel: 40,578 nouns, 24,514 verbs, 86,686 others
      iii. Stein's autobiography: 29,834 nouns, 17,673 verbs, 54,021 others
      iv. Stein's novel: 25,521 nouns, 17,437 verbs, 55,472 others
8. Noun and verbs:
   a. Noun-verb ratios:
      i. Wharton's autobiography: Noun-verb ratio of 1.955
      ii. Wharton's novel: Noun-verb ratio of 1.655
      iii. Stein's autobiography: Noun-verb ratio of 1.688
      iv. Stein's novel: Noun-verb ratio of 1.464
   b. The Chi-square probability for all three comparisons is close to null ($2.28 \cdot 10^{-27}$ for autobiographies, $3.95 \cdot 10^{-22}$ for novels, and $1.4 \cdot 10^{-37}$ for the total works).
9. Plays, years, and words:
   b. Spearman's rank correlation coefficient is 0.927.

8 Network theory and applications in humanities

In Chapter 2, we briefly introduced the concept of a network (commonly referred to as a "graph" in mathematics and a "network" in sociology). Over the next two chapters, we will explore networks more thoroughly, examining both their structure and their application within the humanities. Chapter 8 focuses on the fundamental building blocks of networks – defining key terms, exploring basic concepts, and understanding different types of networks and their behaviors. In contrast, Chapter 9 will shift towards practical analysis, introducing metrics such as centrality, betweenness, and clustering that allow us to measure the influence of nodes, identify key connections, and draw deeper insights from the structure of a network. Together, these chapters provide the foundation and analytical tools for applying network theory to complex humanities data.

8.1 What is a network?

A network consists of nodes (or vertices) that are partially connected by edges. Mathematically, a network is defined as an ordered pair of two sets: a set of nodes V and a set of edges E. Networks can also be represented graphically, as seen in diagrams like Figure 2.7.

Networks represent connections between different entities. In today's world, almost any relationship between objects is described as a network or as part of a larger one. Whether referring to social, biological, geographical, or historical connections, the network perspective remains the same: rather than viewing objects as independent entities, they are viewed as constantly relating to others. In network research, individual entities are rarely the whole story; the connections between them are not only important but crucial for understanding broader patterns and events.

The fundamental approach of network theory differs from that of statistics. Statistics aims to describe an entire dataset using numerical values or verbal descriptions that reflect "general" or "average" behavior. Whether it's central values like mean, median, or mode, measures of dispersion such as variance and skewness, or various correlation coefficients, individual data points lose their significance in favor of the overall dataset and its properties. In contrast, network

DOI: 10.4324/9781003502814-9

theory is just as much about connections between particular individuals, or between individuals and their surroundings as it is about the whole network. As Weingart (2014) explains: "Network analysis in the humanities provides a way to bridge personal stories with the bigger picture ... it allows us to see both the **forest** and the **trees**" (emphasis in the original).

Nevertheless, networks, whether in their mathematical or visual forms, are simply abstract models. While they provide a structured representation, nodes and edges often have properties that extend beyond the purely mathematical aspects of the network. For instance, if we create a network using the set of nodes {Alice Guy, Alfred Hitchcock, François Truffaut, Sergei Eisenstein, Greta Gerwig, *The Birth of Children, The Birds, The 400 Blows, Battleship Potemkin, Barbie*}, where each director is connected by an edge to a film they directed, we end up with the diagram shown in Figure 8.1.

It is important to clarify that, unless stated otherwise, the position of the points or the length of the lines in the visual representation of the network has no inherent meaning. From a "network" perspective, Figure 8.1 contains 10 nodes that are identical in nature and 5 edges that differ only by the labels of the nodes they connect. However, this is merely a model representing real-world objects with additional properties. Each director has a birth date, lived in various locations, holds specific ideologies, and has a distinct filmography. Likewise, each film has a runtime, color scheme, artistic style, plot, and cast. The edges, representing the

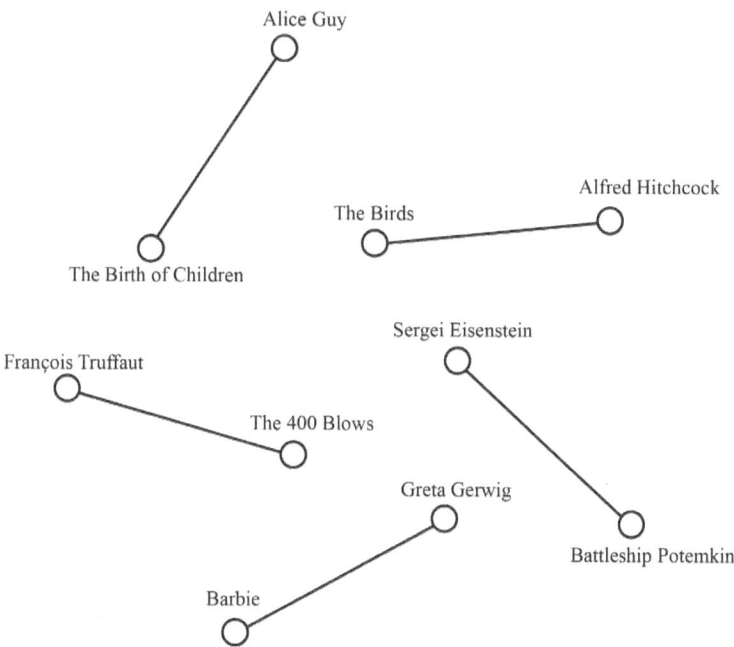

Figure 8.1 A network connecting directors to the films they created.

connection between a director and their film, can be described by factors such as the film's release date, the start of production, or the nature of the relationship between the director and the project (whether they were hired for an ongoing production or initiated the project themselves). The network provides us with a framework for analyzing these relationships (in this case, the connections between directors and their films), but understanding its structure requires additional context, and for this, the properties of the components are essential. The role of digital humanities scholars is to interpret the network in light of the real-world attributes of its components.

8.2 The power of visualization in network analysis

When we use networks, we assume that the structure of the network and the nature of its various connections influence the behavior of the objects within it. For example, Seth Benzell and Kevin Cook (2021) explored how familial ties between European royal houses affected political and military conflicts. They collected data on all family connections among European royal houses from 1495 to 1918 and analyzed this information using network theory. To illustrate their findings, we will present two networks from their study, depicting the situation in 1618.

In 1618, the conflict that would later be known as the Thirty Years' War began. This war was essentially a series of interconnected conflicts that drew in all the European powers. The widespread atrocities committed during this period prompted a significant shift in global diplomacy. Among the war's many outcomes were the decline of Spain's dominance, the rise of French influence, the fragmentation of the Holy Roman Empire into autonomous principalities, and the emergence of the territorial state as the leading political model in the modern world.

Figure 8.2 illustrates familial relations rather than actual political alliances. The left map reveals three clusters: a southern cluster that includes the Catholic monarchs of France, Spain, and southern Italy; an eastern cluster with the Catholic monarchs of Austria, Bohemia, and Poland; and a northern cluster comprising the Protestant monarchs of England, the Netherlands, Denmark, and the German principalities (Prussia, Saxony, and the Palatinate). The map on the right shows that the Austrian Habsburg dynasty (the eastern cluster in the left map) had distant family ties to the southern Catholic cluster in southern Europe. According to Benzell and Cook, these family connections formed the backbone of the two major alliances in the war. Additionally, the intersections of clusters from both maps highlight the centrality of the German principalities in the conflict between the Catholic and Protestant factions, offering insight into how a local rebellion in Bohemia (1618) escalated into a war that reshaped Europe.

It is important to note again that the network is a simplified representation, reducing the system to an abstract structure that reflects only patterns of connections. While this approach has its drawbacks, it also offers significant advantages. One key advantage of network theory is its ability to visually represent networks through diagrams. As demonstrated in Figure 8.2, visualization serves as a

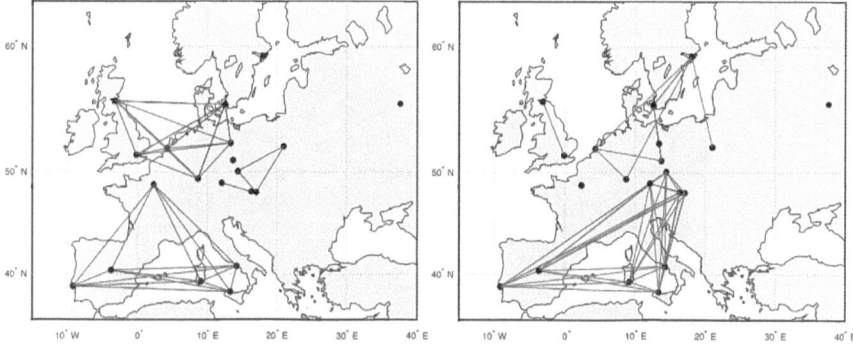

Figure 8.2 Networks of family ties between dynasties in 1618, displayed on maps of Europe. In these maps, the black dots represent capital cities. In the right map, an edge is drawn if two monarchs share a family connection that is at least three generations apart (such as a great-grandfather or great-grandmother). In the left map, an edge exists if two monarchs are connected by direct or indirect family ties among living individuals, including marital relationships.

Source: Benzell and Cooke (2021).

powerful tool in network analysis because it allows us to directly and intuitively grasp the network's structure and properties. However, the effectiveness of this method decreases as the number of nodes and edges increases. When there are too many elements, the visualization can become cluttered and is often referred to as a "hairball".

Consider the following network (Figure 8.3), which illustrates the citation links between 32 American daily newspapers from 1841 to 1884. In this network, each node represents a newspaper issue, and an edge exists between two issues if they contain overlapping sentences. This network is quite extensive, comprising 31,055 nodes and 818,210 edges.

Unlike the smaller network of royal family ties in 1618, it is impossible to extract meaningful insights from a direct visual inspection of the citation network of 19th-century American dailies. To tackle this, network theory offers various metrics that aid in analyzing a network and its properties without depending solely on visual interpretation.

8.3 Basic types of networks: simple, directed, and weighted

Most of the networks we will study in this book have, at most, a single edge between any pair of nodes. In rare cases where multiple edges do exist between the same pair of nodes, these are known as **multiedges**. Similarly, most networks in this book do not have edges that connect a node to itself, although there are some exceptions. These connections are referred to as **self-edges** or **self-loops**. A network that has neither self-edges nor multiedges is called a **simple network**

Figure 8.3 A network of 32 American daily newspapers from 1841 to 1884, where
   nodes represent issues and edges denote shared content through overlapping
   sentences.

or **simple graph**. In contrast, a network that includes multiedges is referred to as
a **multigraph**.

For example, flight connections between countries can be represented by a single
edge to indicate if a flight connection exists. Alternatively, they can be represented
by multiple edges, with each edge depicting a different flight or airline. In this context, a self-loop would represent a domestic flight where the plane takes off and
lands within the same country.

Questions

1. Examine the network in Figure 8.4 and answer the following questions:
    a. *How many edges does this network contain?
    b. *What is the minimum number of edges that needs to be removed to
        make the network simple?

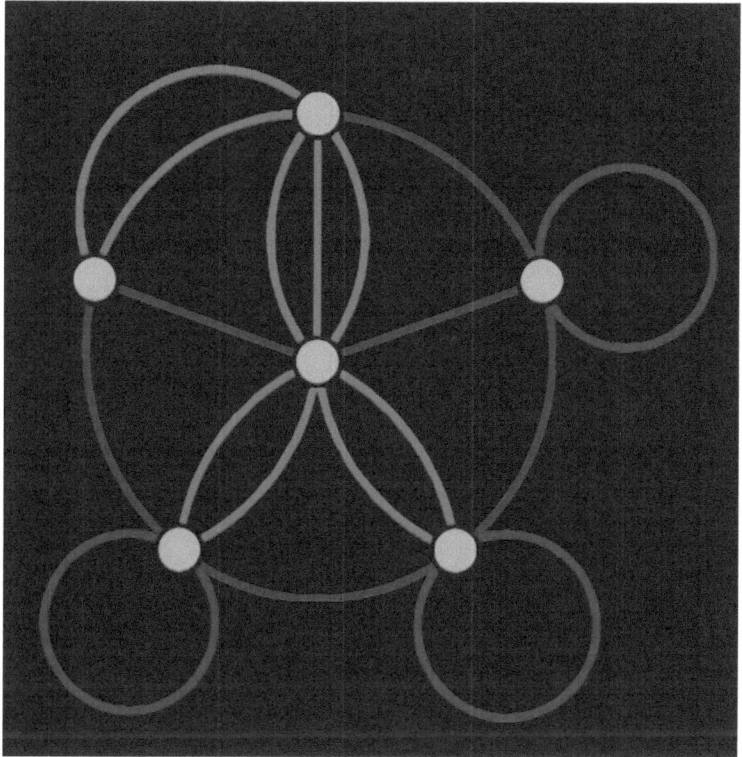

Figure 8.4 A non-simple graph.

Many networks have edges that represent simple on/off connections between nodes – either the edges are present, or they are not. These are known as binary networks. However, in some cases, it is useful to assign strength, weight, or value to edges to quantitatively express the significance of the connection.

For instance, it might be more meaningful to numerically represent family ties between rulers in 1618 based on their degree of relation. Siblings or a parent and child would be represented by a stronger edge than two cousins or the connection between the parents of a married couple. Similarly, links between two daily newspaper issues might have weights that represent the number of overlapping sentences between them.

Another example is a road map, where an edge represents a road connecting one location to another. While this map is a network, representing it as a binary network would prevent us from evaluating the distances between locations, as all distances would be perceived as equal. A classic example of a binary road map is a transit map that illustrates the routes and stations within a public transport system (Figure 8.5). These maps were created to help passengers navigate the system,

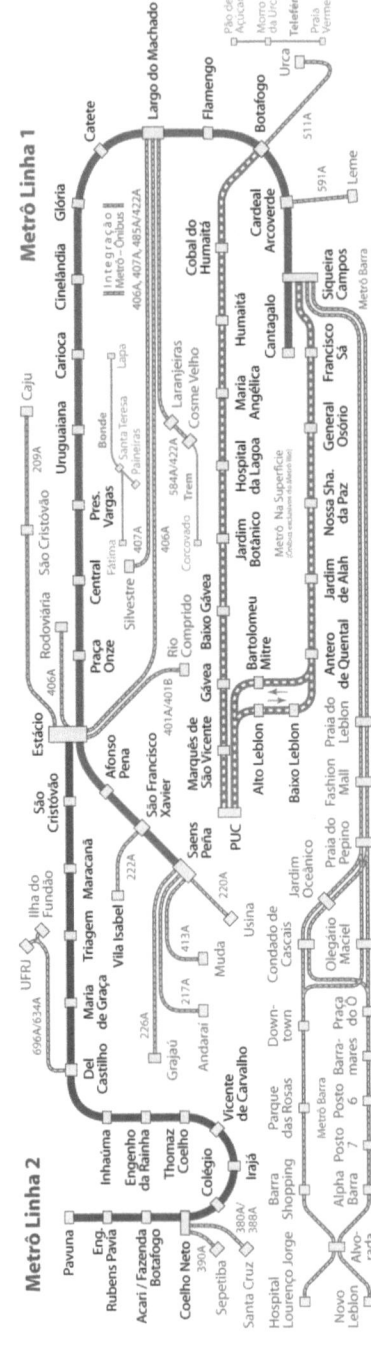

Figure 8.5 A 2007 map of the Rio de Janeiro subway network. Created by Maximilian Dörrbecker.

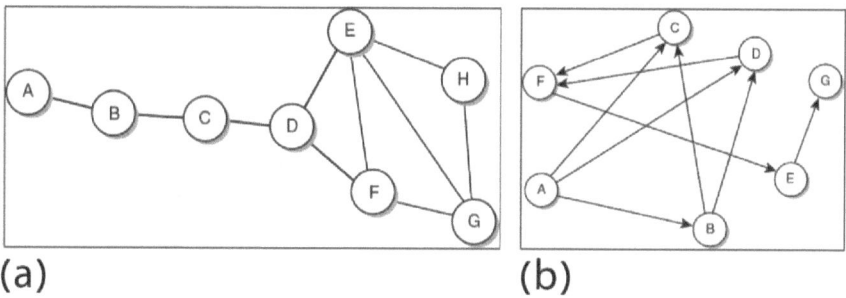

Figure 8.6 An undirected (a) and directed (b) networks.

operating under the assumption that distances between stations or topographical features are not significant. In Figure 8.5, for example, the position of stations and the length of lines hold no practical meaning, as the map reflects only the connections.

A network in which edges have numerical values is called a **weighted network**. These networks can be visually represented by varying the thickness of the edges to indicate different weights (Figure 8.9). While weights are typically positive, there is no reason not to assign negative weights when relevant. For instance, in a social network, positive edges could represent degrees of friendship, while nega- tive edges might indicate levels of hostility.

Just as weights can be assigned to edges, they can also be assigned to nodes, along with other properties such as color or direction (Figure 8.6b). Directionality is particularly significant. A **directed network** is one in which each edge has a specific direction, pointing from one node to another. These edges are known as **directed edges** and are typically represented as arrows. Citation networks, which map the relationships between literary or academic works through citations, are examples of networks with clear directionality.

Questions

2. For each of the following networks, decide whether they should be defined as binary or weighted, directed, or undirected:
   a. A network that examines ancient geography using georeferences in the bible. The nodes represent places mentioned in the bible, and an edge exists if the place are mentioned in the same passage.
   b. A network that examines the lexical division of the Swahili language. The nodes represent words in Swahili, and an edge exists if the words share the same part of speech.

c. A network that examines the relationships between characters in the play *Hamlet*. The nodes represent characters from *Hamlet*, and an edge exists if one character speaks to another.
d. A network that examines the history of cinema through the analysis of citations between films. The nodes represent films, and an edge exists if one film cites another.

In the ancient geography network, the direction of the connection is irrelevant since we're focusing on the joint appearance of settlements in the same passage. However, the frequency of these co-occurrences matters, so this network will be weighted. In the lexical network, neither direction nor weight is relevant because it's a binary question of whether two words share the same part of speech. In the *Hamlet* character network, directionality is crucial, as there is a difference between who is speaking and who is listening. The number of dialogues between characters also matters, making this network both directed and weighted. In the cinema citation network, direction is important because it distinguishes between the citer and the cited. However, the number of citations in a single film isn't particularly significant; simply being cited indicates a film's importance and influence.

Of course, your interpretation of these networks may differ from the ones outlined here. Furthermore, the choice of network type depends on the specific questions we aim to answer.

8.4 Bipartite graphs

Our previous examples involved similar types of entities: places linked to places, words to words, characters to characters, and so on. However, many networks model relationships between two distinct sets of entities. For instance, in a social network, one set could represent people and the other set could represent the groups they belong to. Whenever we connect different types of nodes, we create a **bipartite network**.

In a bipartite network, edges run only between nodes of different types, meaning there are no edges connecting nodes within the same group. An example of a bipartite network is shown in Figure 8.1, where directors are connected to films, but not directors to directors or films to films.

Other examples include connections between people and their places of residence, authors and their works, or objects and their properties. An example of a bipartite network can be found in Giovanna Ceserani's research on English architects in the 18th century (Ceserani 2015). To explore the connections among the architects, Ceserani constructed a network that illustrates the educational backgrounds of English architects (Figure 8.7). This network is bipartite because it connects architects with various types of professional training: practical training as apprentices, training in art schools, education at prestigious English universities

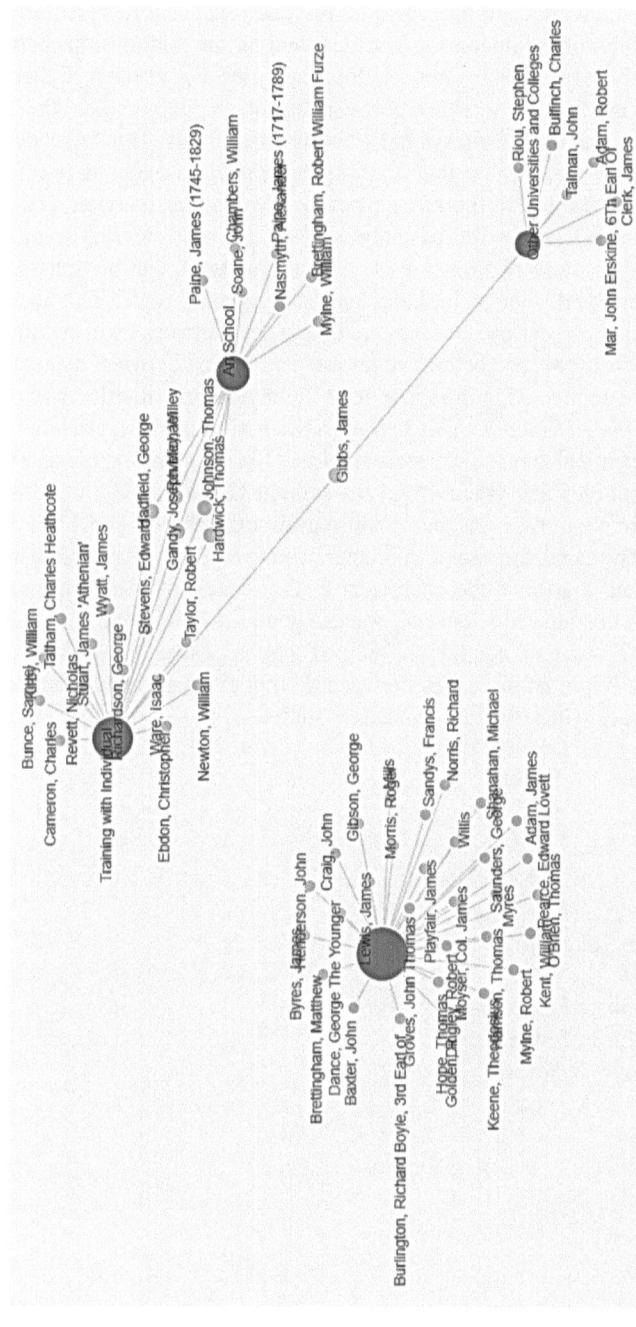

Figure 8.7 A network depicting the educational background of British architects who traveled to Italy. Dark circles in the center represent types of training, while light circles represent the architects. The large cluster in the bottom left corner includes architects with no reliable information on their professional education. http://republicofletters.stanford.edu/publications/grandtour/education/

(Oxford, Cambridge) or law institutions, training in other universities or colleges, and those for whom there is no information available.

While bipartite networks can provide a more comprehensive representation of a system, we are often more interested in understanding the relationships between nodes of the same type rather than between different types. For instance, Ceserani's research focused on the connections between architects, rather than the links between architects and their professional training institutions. Her bipartite network is based on the assumption that a stronger connection exists between two architects who were trained in the same place or in a similar manner. Thus, the bipartite network serves as a means to analyze the relationships among architects.

To further analyze these relationships, a bipartite network can be transformed into a unipartite network, which includes only one type of node. This process, known as a unipartite projection, involves creating a new network by concentrating on one set of nodes. In this projection, edges are established between members of the set if they were connected to the same node in the original bipartite network.

Shakespeare's play *The Comedy of Errors*, written around 1594, tells the story of two pairs of identical twins separated at birth. The comedic series of events begins when Antipholus of Syracuse and his servant Dromio of Syracuse arrive in Ephesus, where their twin brothers, Antipholus of Ephesus and Dromio of Ephesus, reside. By using the list of characters in each scene of the play, a network can be created that links the characters to the scenes in which they appear (Figure 8.8). From this bipartite network, one can generate a weighted network that includes only the characters, connecting them if they appeared in the same scene (Figure 8.9). The weight of the edges between the nodes, then, would depend on the number of scenes which the two characters shared.

Questions

3. Based on Figure 8.9, classify the following characters as either main or minor characters:
    a. Duke of Ephesus
    b. Aegeon
    c. Antipholus of Ephesus
    d. Antipholus of Syracuse
    e. Dromio of Ephesus
    f. Dromio of Syracuse
    g. Adriana
    h. Luciana
    i. Luce
    j. Balthazar
    k. Angelo
    l. Courtezan
    m. Doctor Pinch

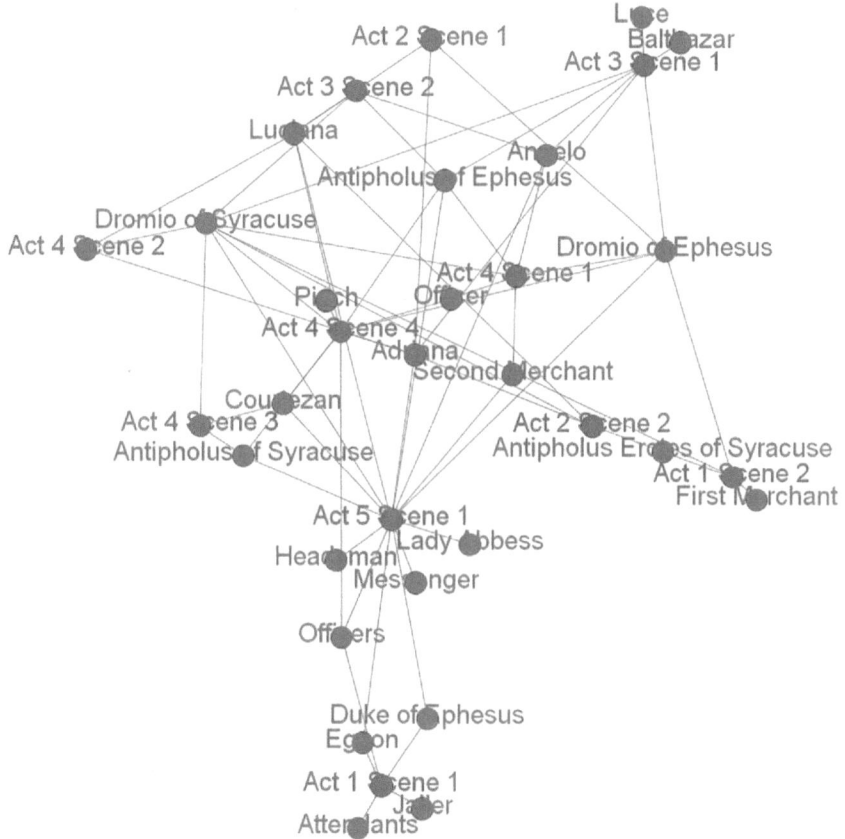

Figure 8.8 Network visualization of Shakespeare's *The Comedy of Errors*, with some nodes representing characters and others representing scenes.

Understanding the hierarchy by analyzing the visual image (Figure 8.9) can be challenging, as we might attribute more significance to the actual placement of nodes than it merits. The position of each node is a constraint of representing the network in a two-dimensional visualization. However, we can observe a cluster of 8 characters in the center of Figure 8.9 who frequently appear together – Antipholus of Ephesus, Antipholus of Syracuse, Dromio of Ephesus, Dromio of Syracuse, Adriana, Luciana, Angelo, and the Courtesan. Other characters, such as the Duke of Ephesus, Aegeon, Luce, Balthazar, and Doctor Pinch, rarely co-appear with others. This division aligns, more or less, with the play's main characters.

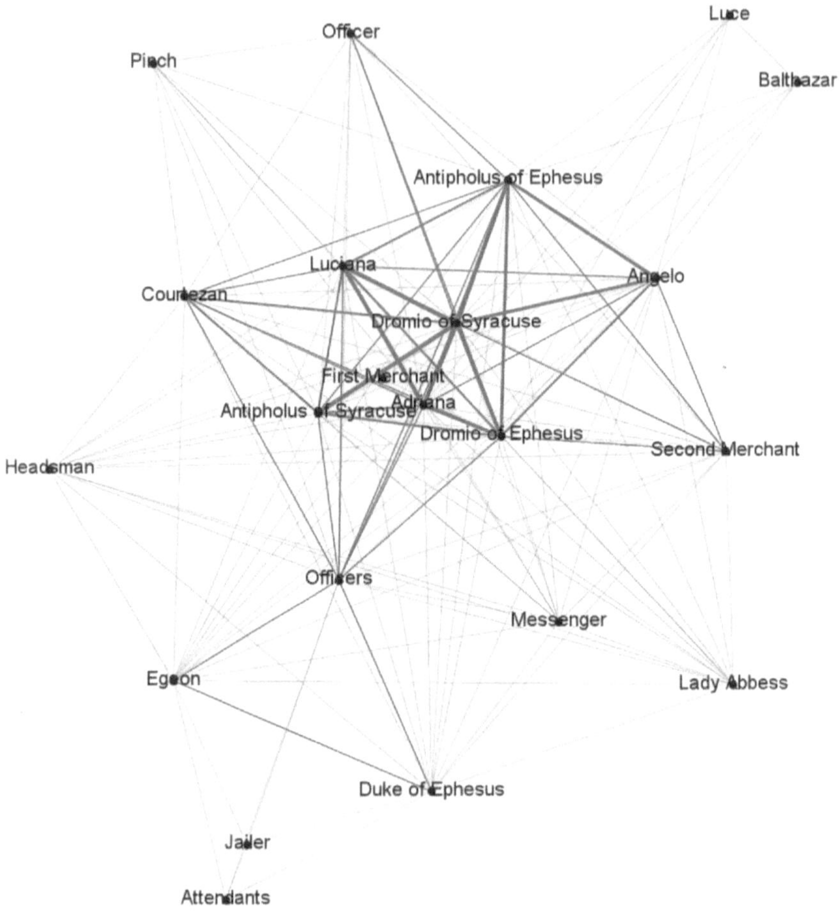

Figure 8.9 Network visualization of the characters in Shakespeare's *The Comedy of Errors*. Edge thickness indicates the number of scenes the characters share.

8.5 Degrees

In network theory, the **degree** of a node is a simple yet highly useful measure for describing a network's structure. It refers to the number of edges connected to a given node, offering insights into that node's level of connectivity within the network. The degree is denoted as $\mathbf{deg}(x)$ or $d(x)$ for any node x. Understanding degrees helps us identify key players, assess influence, and reveal the overall topology of the network.

Questions

4. *Calculate the degrees of each of the eight nodes in the network shown in Figure 8.6a.

The meaning of a degree varies depending on the context of the network. In the network of monarchical familial relations from 1618, the degree of a node represents the number of family ties a monarch has with other European monarchs. In the network depicting the characters in *The Comedy of Errors*, the degree of each node indicates the number of characters that shared scenes with a particular character (see Table 8.1). Additionally, one can define a **weighted degree**, which sums the weights of all edges connected to a node, rather than simply counting the number of edges (see Table 8.1).

Although this is merely the count of shared appearances in scenes, it provides a general impression of the centrality of the two servants (Dromio of Syracuse and Dromio of Ephesus), their masters (Antipholus of Ephesus and Antipholus of Syracuse), Adriana (the wife of Antipholus of Ephesus), and Luciana (her sister). This metric offers a more reliable measure of centrality than the visual impression discussed in the previous section.

Table 8.1 Degrees and weighted degrees of the characters in *The Comedy of Errors*

Character	Degree	Weighted degree
Dromio of Syracuse	19	47
Dromio of Ephesus	19	39
Antipholus of Ephesus	18	37
Adriana	18	36
Luciana	16	33
Antipholus of Syracuse	17	31
Angelo	17	28
Officers	18	27
Courtesan	16	25
Second Merchant	15	19
Duke of Ephesus	16	18
Aegeon	16	18
Officer	11	14
Lady Abbess	14	14
Headsman	14	14
Messenger	14	14
Pinch	9	9
Balthazar	6	6
Luce	6	6
Jailer	4	4
Attendants	4	4
First Merchant	3	3

In directed networks, it is common to define both indegrees and outdegrees. The indegree measures the number of edges entering a node, while the outdegree measures the number of edges exiting a node.

Questions

5. Consider Figure 8.6b:
    a. *Calculate the indegrees of the nodes.
    b. *Calculate the outdegrees of the nodes.

8.6 Paths, connectivity, and connected components

A network is more than just a static framework that maps the objects involved (the nodes). It can also be seen as a dynamic space, whether it's people traveling between cities, messages moving through a social network, or information flowing across a computer network. In network theory, a **path** is simply a way to travel from one point to another by following the connections. Imagine a network as a map of cities (the points) connected by roads (the connections). A path is like the route you would take to travel from one city to another.

Mathematically, a path in a network is defined as any sequence of nodes where each consecutive pair of nodes in the sequence is connected by an edge. In directed networks, movement along an edge is only possible in the direction specified by the edge. In undirected networks, movement can occur in either direction along the edges. Typically, an additional condition is included in the definition of a path: no edge can be repeated, although a node may be revisited multiple times. A **cycle** is a special type of path that begins and ends at the same node.

The length of a path is defined as the number of edges it includes, not the number of nodes. This definition aligns with the usual concept of length, which measures the distance between points rather than the points themselves. A pair of nodes in a network will typically be connected by many paths of many different lengths. The **distance** between two nodes is considered the length of the shortest path between them. If no path exists between two nodes, they are said to be disconnected, and the distance between them is considered infinite, a convention used for clarity. The **diameter** of a network is the greatest distance between any two nodes within the graph.

Questions

6. Consider Figure 8.6a:
    a. *What is the distance between the following pairs of nodes (A,F), (H,C), (F,E)

b. *What is the diameter of the network?
c. *For which pairs of nodes is the distance between them equal to the network's diameter?

Paths are crucial for identifying indirect connections within a network. For example, Renzhiwula Marianne Roxas and Giovanni Tapang (2010) used networks to automatically differentiate between prose and poetry. They constructed networks using the set of words in various texts, with an edge connecting two nodes if the words were adjacent in the text. Their research, based on 60 works categorized as poetry and 34 as prose (all sourced from the Gutenberg Project), demonstrated that in prose, the distances between pairs of words and the network's diameter tend to be shorter due to the frequent repetition of words. In contrast, the use of unique words in poetry leads to longer distances and a larger network diameter.

Paths are not inherent properties of networks; they are objects that exist within a network, much like nodes and edges. However, paths allow us to define an important network property – **connectivity**. A network is considered connected if there is a path between every pair of nodes. For instance, the network of characters in Figure 8.9 is connected because a path exists between every pair of characters. In contrast, Figure 8.2 shows no paths between the Catholic capitals of southern Europe and the Protestant capitals of northern Europe, indicating a disconnected system. This lack of connectivity forms the basis for the researchers' conclusion about the two opposing alliances in Europe in 1618.

The subgroups identified in Figure 8.2 are known as connected components. A **connected component** is a subset of nodes for which a path exists between any two nodes and there is no additional node in the network adjacent to the nodes in this subset. A connected network consists of a single connected component, whereas a disconnected network contains multiple connected components. Referring back to the previous examples: Figure 8.1 has 5 components, Figure 8.2 has 3 components (in both the left and right maps), Figure 8.5 has a single component, and Figure 8.7 contains two components.

In a 2011 paper, Franco Moretti analyzed the play *Hamlet* using network theory (Moretti 2011). Two characters in *Hamlet* were connected by an edge if they both spoke in the same scene (Figure 8.10). To simplify the network's complexity, these edges are binary (unweighted) and undirected. One of Moretti's claims in his paper is that Hamlet, the play's tragic protagonist, and his close friend Horatio play crucial roles as bridges that hold the social world of the play together, preventing its disintegration. The network of the play is connected, as shown in Figure 8.10. However, when Hamlet and Horatio are removed (Figure 8.11), the network becomes disconnected. According to Moretti, the significance of these two characters lies in their unique position as the only ones linking the royal court (with its intrigues), depicted on the left of Figure 8.11, to the civilian world – the realm of ambassadors, messengers, and sailors.

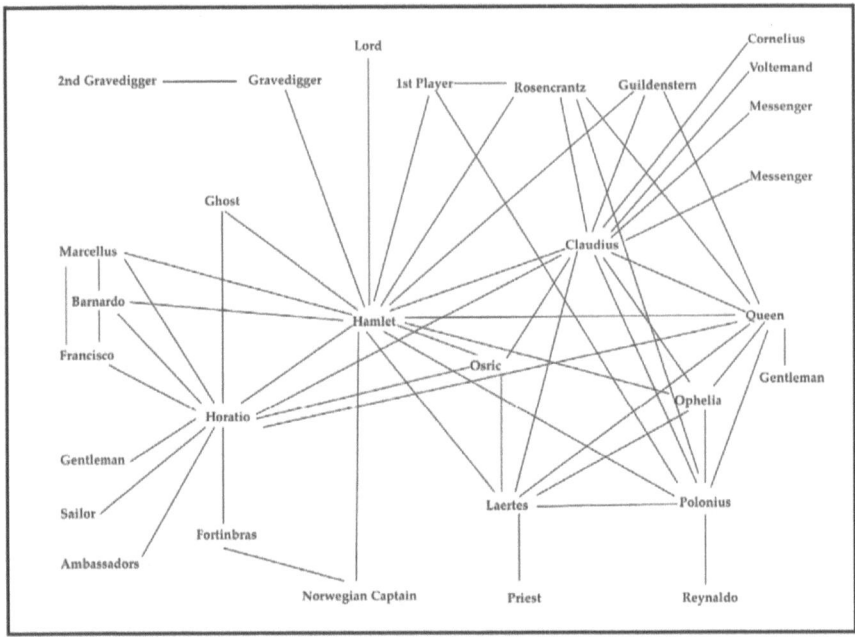

Figure 8.10 Network of connections between characters in Shakespeare's *Hamlet*.
Characters are connected if they speak in the same scene.

Source: Moretti (2011).

8.7 Trees and spanning trees

Trees are a fundamental type of network, characterized by their simplicity and structure. A tree is a connected network with no cycles, making it an ideal model for hierarchical relationships or branching structures. In simple terms, a tree resembles an actual tree with roots, branches, and leaves. Imagine starting with a single point, representing the "root" of the tree. From this root, more points branch out, and each of these points can further branch into additional points.

A simple example of a tree network is a family tree, assuming no intra-familial marriages. In a family tree, each person (node) is connected to their descendants (children) and ancestors (parents) in a hierarchical structure. Trees are commonly used in everyday life to organize information in a clear and straightforward manner, such as organizing files on a computer, mapping family relationships, or illustrating how decisions lead to different outcomes. A river system is another example of a real-world tree-like network, as it is connected and free of cycles. In computer science, trees are essential in data structures, enabling efficient searching, sorting, and data management.

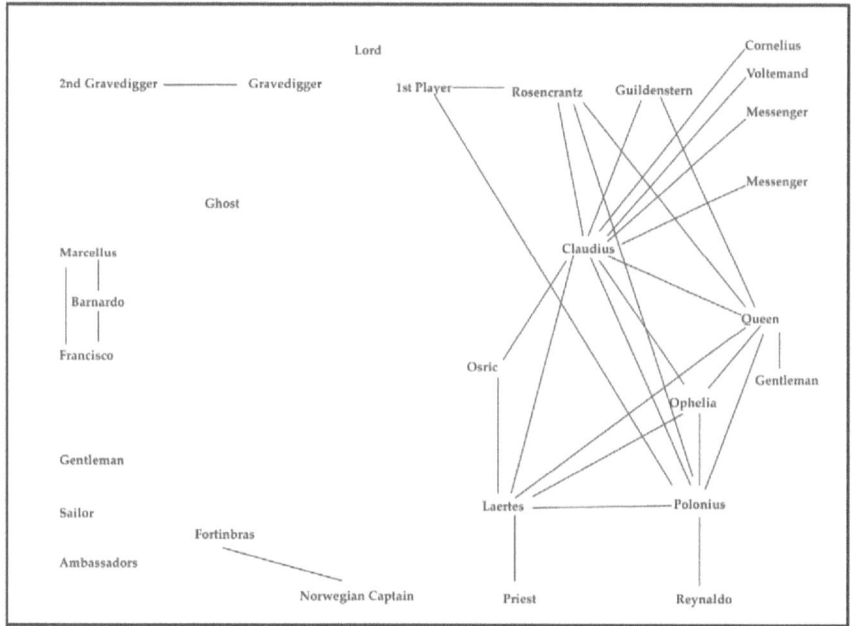

Figure 8.11 Network of connections in *Hamlet* with Hamlet and Horatio removed.

Source: Moretti (2011).

Trees have several important properties that make them useful in the study of complex networks. First, because a tree has no cycles, there is exactly one path between any pair of nodes. This stems from the fact that if there were two different paths between nodes A and B, you could travel from A to B along one path and return along the other, creating a cycle. This would contradict the definition of a tree, which must be free of cycles.

Building on this idea, in any connected network, there exists a subgraph that includes all the nodes connected by the minimum number of edges required to link them without forming loops or cycles. This subgraph is called a spanning tree. A **spanning tree** can be metaphorically described as the skeleton of the original network. Spanning trees are crucial because they help identify the simplest and most efficient way to connect all elements in a network, such as when designing road systems, electrical wiring, or computer networks, without including any unnecessary paths.

For example, Figure 8.12a is not a tree because it contains cycles. By removing edges from these cycles, we can create a cycle-free network. Figures 8.12b and 8.12c show two possible spanning trees. While a network can have multiple spanning trees, each one offers an efficient, yet reductive, way to visualize the network, without losing its connectivity.

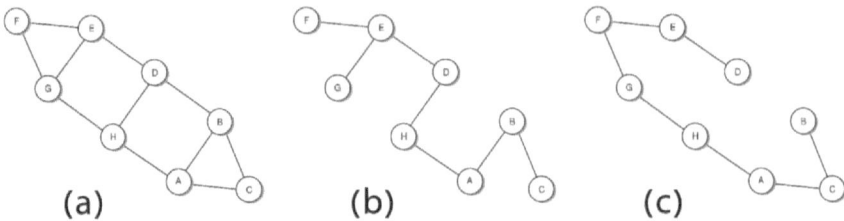

Figure 8.12 A connected network (left) and two potential spanning trees.

8.8 Building and analyzing networks in Python

In this section, we will apply the concepts learned so far by building and analyzing networks using Python. This hands-on exercise will introduce you to fundamental techniques for creating network structures, adding vertices and edges, and performing basic network analysis.

To create a network in Python, we start by importing the `networkx` library:

```
import networkx as nx
```

Let's take a moment to discuss how the import function is used here. In Chapter 7, we imported the `nltk` library and prefixed its functions with the library name, like so: `tokens= nltk.word_tokenize(rawtext)`. In the code box above, we use `as` to create a shorthand for `networkx`, allowing us to use nx as a prefix. For example, we'll use `nx.Graph()` instead of `networkx.Graph()`. Both approaches are equivalent but using the shorthand makes the code more concise.

The first step in creating a network is to call the `Graph` function from the `networkx` library (or `nx` for short). This function returns a graph-type variable with various methods available for manipulating the network:

- `add_node()`: Takes the name of the node (number or string) as an argument.
- `add_edge()`: Takes a pair of nodes as an argument and adds an edge between them. If the nodes do not already exist, this method will create them.

Here's an example of how to create a network with two nodes, "A" and "B", connected by an edge:

```
G = nx.Graph()
G.add_node("A")
G.add_node("B")
G.add_edge("A","B")
```

To visualize the network, we can use the `draw` method. This method requires two arguments: the graph variable (`G` in this case) and an instruction to display the labels of the nodes:

```
nx.draw(G, with_labels=True)
```

While `add_node` and `add_edge` are useful for adding individual nodes or edges, they are limited to adding these elements one at a time. This is not very convenient when our graph includes many nodes and edges. For this purpose, the methods `add_nodes_from` and `add_edges_from` allow us to pass a list of nodes or edges to be added to the graph at once.

```
G.add_nodes_from(["A", "B", "C"])
G.add_edges_from([("A","B"), ("A","C")])
```

To further explore different properties of the network, we can use the methods `number_of_nodes()`, `number_of_edges()`, and `degree()`. For instance, the following code will print the number of nodes and edges in the graph:

```
print("Nodes ", G.number_of_nodes())
print("Edges ", G.number_of_edges())
```

Questions

7. *Write a program that creates the network shown in Figure 8.6a and draws it with labels.

The `degree` method can be used in two ways: calling it with a specific node as an argument returns the degree of that node, while calling `degree()` without any arguments returns a list of pairs, with each pair consisting of a node and its degree.

Questions

8. *Predict the output of the following code snippets for the network in Figure 8.6a.
   a. G.degree("A")
   b. G.degree(["A","C"])
   c. G.degree()

In section 2.3.3 we learned the Breadth-first search algorithm, which can be used to find the shortest path between two nodes in a graph. The Python function for finding the shortest path is `shortest_path()`. This function requires the graph variable, a source node, and a target node as arguments, returning the path as a list of nodes.

Questions

9. Paths and distances:
   a. *What will be the result of the following code:

```
path = nx.shortest_path(G, source="D", target="H")
```

   b. Which Python function could you use to measure the distance between nodes D and H?

In the previous chapters, we learned that `len()` can measure the length of lists. Therefore, `len(path)` will count the number of nodes in the shortest path between D and H. However, since distance is defined as the number of edges, not nodes, the correct distance would be `len(path) - 1`.

The `diameter()` function computes the maximum distance between any two nodes in the network. For example, `nx.diameter(G)` will return 5, which is the longest distance in this graph (as calculated in Question 6).

Glossary

Bipartite Network A type of network where nodes are divided into two sets, and edges only exist between nodes from different sets, not within the same set.
Centrality A measure of the importance of a node in a network.
Connected Component A subset of a network in which all nodes are connected to each other, but no node is connected to nodes outside the subset.

Degree The number of edges connected to a node in a network. It is used to assess how connected a node is within the network.

Directed Network A network in which the edges have a direction, meaning they go from one node to another in a specific order. These are often represented by arrows.

Edge The connection or relationship between two nodes in a network. It can represent various kinds of relationships, such as family ties, citations, or shared content.

Graph Another term for a network, especially in mathematics, consisting of nodes (vertices) and edges that connect them.

Multigraph A network that allows multiple edges between the same pair of nodes, representing different types of connections or multiple interactions between the same entities.

Network A collection of nodes connected by edges, used to model relationships between different entities, such as people, places, or objects.

Node (Vertex) An individual entity in a network, which can represent anything from a person in a social network to a word in a text network.

Path A sequence of nodes connected by edges. A path shows the route one can take from one node to another within the network.

Simple Network A network with no self-edges or multiedges. Each pair of nodes has at most one edge between them, and no node is connected to itself.

Spanning Tree A subgraph that includes all the nodes of a network, connected by the minimum number of edges required to link them without forming any cycles.

Tree A type of network that is connected and free of cycles, often used to represent hierarchical structures.

Weighted Network A network in which edges have numerical values (weights) assigned to them, indicating the strength or significance of the connections between nodes.

References

Benzell, Seth G., and Kevin Cooke. 2021. "A Network of Thrones: Kinship and Conflict in Europe, 1495–1918". *American Economic Journal: Applied Economics* 13(3): 102–133. https://doi.org/10.1257/app.20180521

Ceserani, Giovanna. 2015. "British Architects on the Grand Tour in Eighteenth-Century Italy". http://republicofletters.stanford.edu/publications/grandtour/.

Moretti, Franco. 2011. "Network Theory, Plot Analysis". *New Left Review* 68. https://newlef treview.org/issues/ii68/articles/franco-moretti-network-theory-plot-analysis..

Roxas, Ranzivelle Marianne, and Giovanni Tapang. 2010. "Prose and Poetry Classification and Boundary Detection Using Word Adjacency Network Analysis". *International Journal of Modern Physics* 21(4): 503–512. https://doi.org/10.1142/S012918311 0015257

Weingart, Scott B. 2014. "Networked Society". September 19. https://medium.com/@ scott_bot/networked-society-1120b8ae2704.

Solutions to selected questions

1. Re. Figure 8.4:
   a. The network has 18 edges.
   b. Three loops and 5 double edges need to be removed.
4. The degrees are as follows: $d(A) = 1$, $d(B) = d(C) = 2$, $d(D) = 3$, $d(E) = 4$, $d(F) = d(G) = 3$, $d(H) = 2$.
5. Re. Figure 8.6b:
   a. The indegrees are as follows: $d(A) = 0$, $d(B) = 1$, $d(C) = 2$, $d(D) = 2$, $d(E) = 1$, $d(F) = 2$, $d(G) = 1$
   b. The outdegrees are as follows: $d(A) = 3$, $d(B) = 2$, $d(D) = 1$, $d(E) = 1$, $d(F) = 1$, $d(G) = 0$.
6. Re. Figure 8.6a:
   a. The distances are as follows: $disr(A, F) = 4$, $dist(G, B) = 4$, $dist(F, E)$.
   b. The diameter of the network is 5 (the distance between A and G).
   c. There are two pairs of nodes with a distance of 5: (A,G) and (A,H).
7. Creating a network:

```
G = nx.Graph()
G.add_nodes_from(["A",  "B",  "C",  "D",  "E",  "F",
"G", "H"])
G.add_edges_from([("A","B"), ("B", "C"), ("C", "D"),
("D", "E"), ("E", "F"), ("F", "G"), ("G", "H"),
("G", "E"), ("F", "E"), ("D", "F"), ("E", "H")])
nx.draw(G, with_labels=True)
```

8. Code output:
   a. The code will return the degree of A, which is 1.
   b. The code will return a list of two pairs, [('A', 1), ('C', 2)].
   c. The code will return a list of 8 pairs representing the degrees of all nodes.
9. The result will be the shortest path connecting D and H: ['D', 'E', 'H'].

9 Center and periphery in network theory

As we have already noted, the basic premise of network research is that knowing the structure of a network is essential for understanding various characteristics of the system and its participants. To this end, mathematicians, computer scientists, and social scientists have developed various measures and computational tools. In this chapter we will focus on measures that help understand the relationship between center and periphery as well as connections between subgroups of the entire system.

9.1 Building a network

9.1.1 The E-road network

On March 28, 1947, two years after the end of World War II and at the beginning of the Cold War between the United States and the Soviet Union, the United Nations established the United Nations Economic Commission for Europe (UNECE). The main purpose of the Commission was and still is to encourage and facilitate economic cooperation between European countries (Stinsky 2021). Although the commission did not succeed in its main mission due to the split created in Europe during the Cold War, one of its first and foremost achievements was the creation of the International E-road Network, approved by the United Nations in September 1950. Paul Hartmann Charguéraud, the Director of the UNECE Secretariat's Transport Division, described it this way: "The Working Party on Highways is concerned with no less a problem than that of developing for Europe roads that will serve the needs of our times at least as well as the Roman roads served in ancient times" (Stinsky 2021, 41). Indeed, the system that was designed in 1950 to connect Europe from Scandinavia to Sicily is still used today as a framework for coordinating transport systems in different countries. In fact, in 1989, immediately after the fall of the Berlin Wall, there was already a highway connecting Lisbon in Western Europe to Moscow in the East. Currently, this system also connects Central Asian countries, including Armenia, Azerbaijan, Uzbekistan, Tajikistan, Kazakhstan, and Kyrgyzstan. The longest road is the E40, which stretches for 8,690 km, from the

DOI: 10.4324/9781003502814-10

Figure 9.1 The E-road network (http://en.wikipedia.org/wiki/Image:International_E_R oad_Network.png).

port city of Calais in western France to the Kazakh city Ridder on the Kazakh-China border. Figure 9.1 shows the network of E-roads, as of 2020.

In this chapter, we will use this road system, or rather, the manifested network that connects European cities, as a basis for exemplifying various concepts in network theory. As a preliminary step, we will learn how to create the network itself. We have already learnt how to "manually" create small networks, yet when it comes to large networks, such as the European E-road network, this becomes very laborious. Consequently, a more automated method, based on data files, is preferable.[1]

9.1.2 Creating a network from external data

The information required to create a network is stored in two CSV (comma-separated values) files, which can be found at https://zefsegal.com/computational-literacy-for-the-humanities/):

- A list of nodes (`node_file` below)
- A list of edges (`edge_file` below)

The nodes file contains rows, each row with a name of a node. In the edges file there are two node names separated by a comma – a row for any two nodes connected by an edge. The first row in both files contains titles. When CSV files are opened as spreadsheets (e.g., Excel or Google sheets) the comma is interpreted as a column separator.

When creating a network from data in CSV files we start by importing two modules: `networkx` for the creation of networks, and `csv` for reading tabular data in CSV format.

```
import networkx as nx
import csv
```

Reading the nodes file

In previous chapters, we read the contents of a text file using the function `read()`. This procedure is slightly different with CSV files. In order to read such files we use the method `reader()` included in the CSV module. The variable `node_file` contains the name of the file and its path.

```
# open the file for reading
nodecsv = open(node_file,"r")
# read the contents
csv_content = csv.reader(nodecsv)
```

When a CSV file is read by `reader()`, each of its rows is saved a list of strings. The output (`csv_content`) is not a list per se, so we will use the `list()` function to convert it to a list and then iterate through all its members, and copy each one to a real list of nodes using `append()`. Since the first row of `csv_con-tent` is the title, we will skip it and begin at index `[1]`.

```
nodes = []
for row in list(csv_content)[1:]:
    nodes.append(row)
nodecsv.close()
```

Because the nodes file has only one value in each row, each element in `csv_con-tent` will be a one-member list. Consequently, `nodes` will have the following structure:

```
[['C'], ['D'], ['E'], ['B'],...]
```

As we learned in Chapter 8, the method `add_nodes_from()` that is used for creating nodes in a network requires a simple list of nodes as input. The following code iterates through our previous list and creates a corresponding `node_names` list, in which only the first element of each "internal" list is kept.

```
node_names = []
for item in nodes:
    node_names.append(item[0])
```

Consequently, `node_names` will be a simple list of strings (i.e., node names), which is the appropriate structure to serve as argument for `add_nodes_from()`.

```
['C', 'D', 'E', 'B',...]
```

Reading the edges file

The procedure for defining the edges is similar, with one crucial difference: the edges file consists of the names of two nodes, separated by a comma. As a result, when we convert `csv_content` to a list the result is a list of two-member lists.

```
[['B', 'C'], ['C', 'E'], ['E', 'B'],...]
```

However, since `add_edges_from()` requires a list of tuples, we will need to perform an extra step. As we iterate through the list format of `csv_content`, we will first use the function `tuple()` to convert each "internal" list into a tuple, and then append this tuple to our list of `edges`.

```
for row in list(csv_content)[1:]:
    t = tuple(row)
    edges.append(t)
nodecsv.close()
```

The result – a list of tuples – will serve as the appropriate argument for `add_edges_from()`.

```
[('B', 'C'), ('C', 'E'), ('E', 'B'),...]
```

Questions

1) *Write a program that creates the E-road network (data available at https://zefsegal.com/computational-literacy-for-the-humanities/).
2) *Write a program that prompts a user for two names of cities and returns the shortest path between those two cities and its length. What types of input could cause an error?

Our newly formed network consists of 1,174 nodes and 1,417 edges. We can visualize the network using nx.draw(). However, unlike the network from Chapter 8, the E-roads network is too large for an appropriate visualization. In the following sections we will learn various quantitative measures, that provide valuable information about networks, without requiring their visual representations.

9.2 Centrality measures

A significant part of the research in network theory focuses on concepts of **centrality**. These studies answer the question "Which are the most significant nodes in the network?" by providing appropriate quantitative measures. As we will now show, there is no one single answer to this question, but rather, many answers, depending on the particular definition of centrality. Here are some examples.

Is Hamlet the main character in the play that bears his name because he interacts with more characters than any other character in the play, or because he is the character who connects other characters? (Moretti 2011) In this case, it is relatively easy to claim that he is the main character, in part, due to the name of the play.

What about the centrality of Grigori Rasputin (1869–1916), the Russian mystic who was close to the Russian royal court, and some claim that the scandals associated with him were among the causes of the fall of the royal house in 1917? (Kilcoyne 1961). Rasputin, unlike Hamlet, was not close to many figures, only to the empress Alexandra. His centrality stemmed from his influence on one other major figure.

Switching to a completely different domain, one may ask what made the blockade of the Suez Canal in March 2021 an international media concern (BBC 2021). The importance of the canal stems from the fact that around 12% of all global maritime trade passes through it. Consequently, the economic damage of the blockade was about US$400 million per hour. In other words, the centrality of the canal derives from its being a midpoint of many routes connecting world ports.

Like the Suez Canal, the importance of US Presidents Jimmy Carter, Bill Clinton, and Donald Trump in the peace agreements between Israel and surrounding Arab countries stemmed from them being necessary intermediaries between the adversaries. They too were nodes that stood on the path between other nodes.

All these are examples of key nodes in networks. However, the definition of what constitutes the center was different in each and every example.

9.2.1 Degree centrality

The simplest measure of centrality is the **degree** of a node. The use of this measure is quite intuitive, as the degree reflects the number of links, and accordingly may reflect a level of influence or prestige. For example, the number of followers on Facebook, Instagram, Twitter, or TikTok is the accepted measure for identifying key figures in these networks. In the previous chapter we used this measure to determine the centrality of Dromio of Ephesus and Dromio of Syracuse, the twin brothers in Shakespeare's play *The Comedy of Errors*.

Similarly, in directed networks, the in-degree or out-degree of a node can be used as a measure of centrality. For example, studies in the history of ideas often use citations as a measure of the effect of one text on another. The number of times a particular text is mentioned in other texts expresses its centrality in the textual network. Since this is a directed network (from the quoted text to the quoting text), it is customary to refer to the outgoing degree. This is evident in Google Scholar, a search engine that indexes scholarly literature. The search engine retrieves the relevant articles, along with the number of times they were cited by other publications (i.e., their out-degree). For example, the most cited article in 2020 was "Deep Residual Learning for Image Recognition", an article on artificial intelligence authored by a team of researchers at Microsoft (He et al. 2016). This article had 49,301 citations in 2020 (and 239,504 by October 2024).

Questions

3) *Observe the map of the Hankyu railway lines in the Northern Kansai region of Japan (as of 2022) in Figure 9.2. Which of the following stations are central according to the degree centrality measure (more than one answer may be correct):
    a. Jūsō
    b. Kawanishi-Noseguchi
    c. Takarazuka
    d. Nishinomiya-kitaguchi
    e. Kawaramachi
    f. Awaji

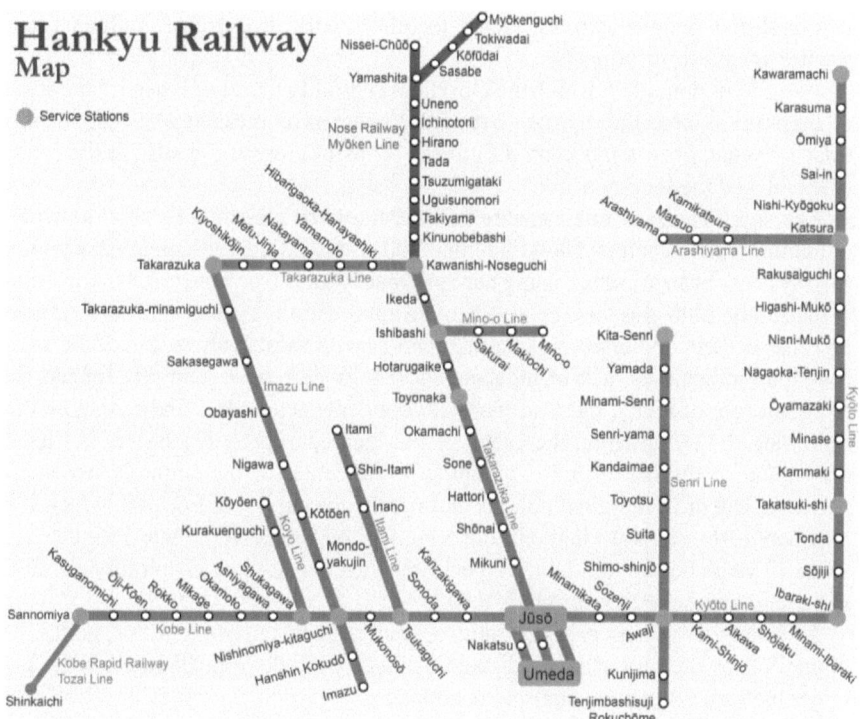

Figure 9.2 The Hankyu railway network (https://commons.wikimedia.org/wiki/File:Han kyu_map.svg).

9.2.2 Closeness centrality

Unlike degree centrality, **closeness centrality** does not rely only on the number of nodes adjacent to a given node but also measures the average distance from the node to all other nodes in the network. According to this measure, nodes are considered more central if they are closer on average to the rest of the nodes. The first to conceive of this measure was the psychologist Alex Bavelas in his 1950 article on patterns of communication and transfer of knowledge (Bavelas 1950). According to Bavelas, "the findings suggested that the individual occupying the most central position in a pattern was most likely to be recognized as the leader".

Between 1420 and 1924, the emperors of China lived in the Forbidden City in Beijing. At that time, it was forbidden for people to enter or leave the city without the permission of the emperors. This greatly reduced the number of possible connections of the emperors themselves, and thus, by degree centrality, the emperors of China were not central in Chinese society. Nevertheless, an alternative measure, closeness centrality, measures the position of a node in relation to the complete network, and not to its immediate neighbors. Accordingly, the rulers of China, as entities located at the center of the Chinese society network of contacts, exhibit a relatively high closeness centrality. In general, it can be assumed that people who are closer to others will have greater influence and greater access to information.

A different domain in which the closeness centrality measure is used is the study of literature, where the distance between characters is measured in units of plot (place, events, or time). A central figure in terms of closeness is also at the center of the plot of the literary work.

The significance of this measure is different when examining spatial networks of administrative centers. David Jenkins (2001) studied the settlement patterns and road systems of Inca culture using network theory tools (see Figure 9.3). According to the author, high closeness centrality of an administrative center reflects an impact on other centers and effective communication with those centers, but at the same time may indicate its lack of independence. Low closeness centrality reflects the marginal role of a city, but also its relative independence. His findings show that according to this measure, the center of the Inca administrative system is Cuzco, the capital of the Inca Empire, which is at the center of the empire's settlement network. The most marginal point is Quito, now the capital of Ecuador, which was conquered from a rival kingdom, the Kingdom of Quito, in the late 15th century, some 40 years before the fall of the empire. Cuzco's closeness centrality is 0.206, and Quito's closeness centrality is 0.093.

To calculate the closeness centrality of each node in a network with N nodes, we divide N-1 by the sum of all the distances of a given node from all other nodes. The larger the result, the more central the node.

We will illustrate this with a simple network consisting of four nodes: 1, 2, 3, and 4. The degree of 1 and 4 is 2, and the degree of 2 and 3 is 3. (See Figure 9.4.)

The distance of 1 from the two nodes 2 and 3 is 1, and the distance between it and 4 is 2, so the sum of the distances is $1 + 1 + 2 = 4$. Therefore, the closeness centrality of node 1 is $\dfrac{4-1}{4} = \dfrac{3}{4}$.

Questions

4) *Calculate the closeness centrality of nodes (2), (3), and (4).

The closeness centrality measure is always between 0 and 1. The value 1 is obtained when a given node is directly connected to all other nodes, like nodes 2 and 3 in the previous example.

This measure has two main shortcomings. First, the values in a given network tend to be close to each other, so it is relatively difficult to distinguish between

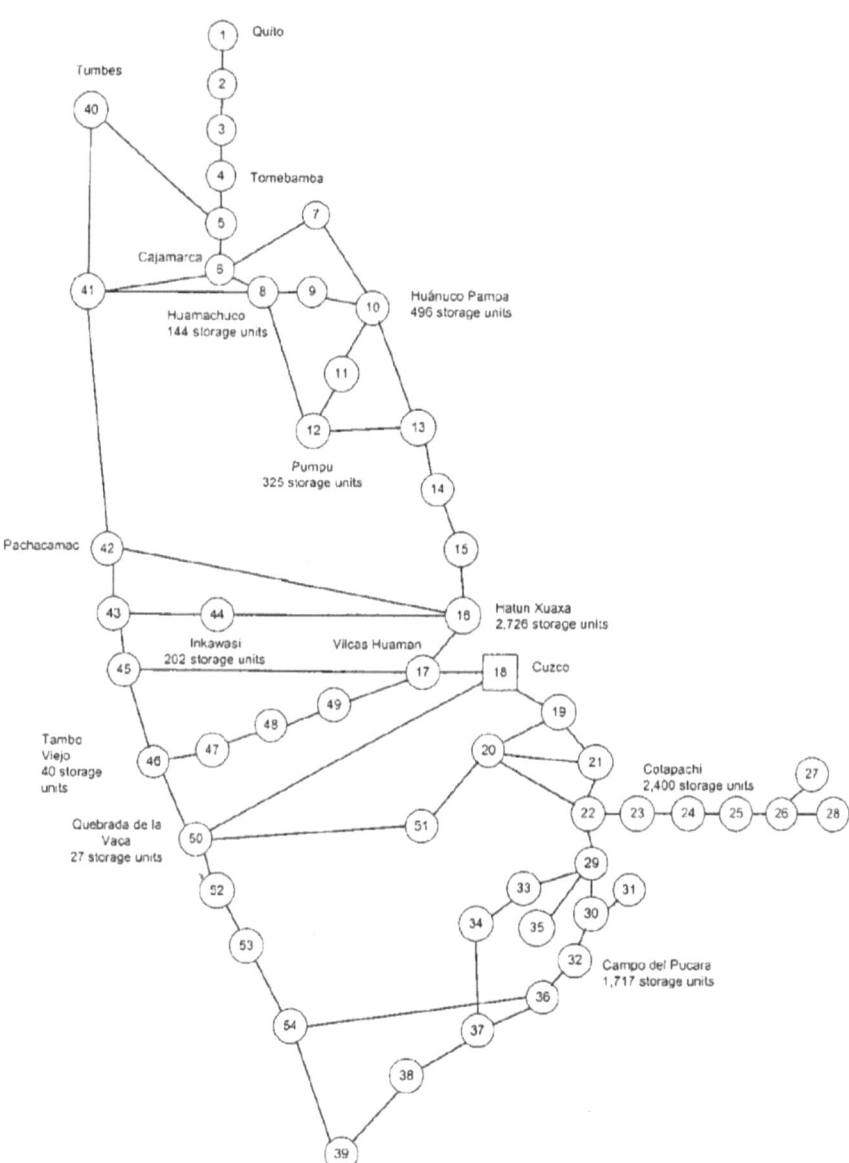

Figure 9.3 The Inca Road network.

Source: Jenkins (2001).

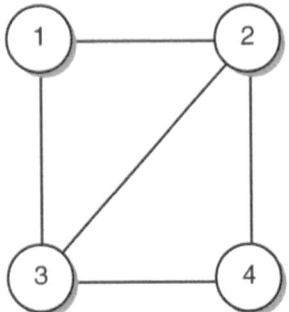

Figure 9.4 A simple network.

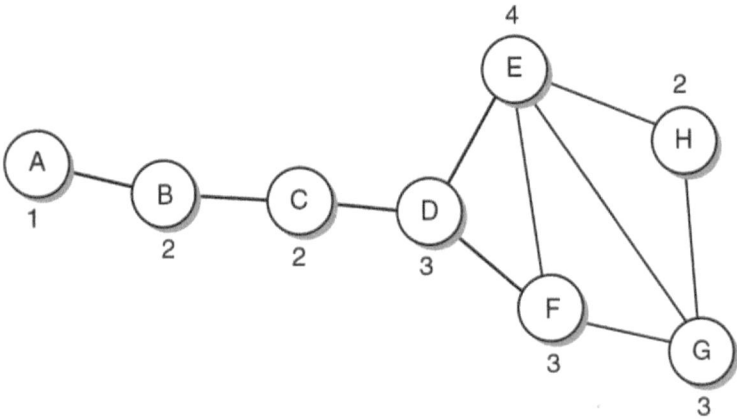

Figure 9.5 Calculating degree centrality.

different levels of centrality. Second, the closeness centrality measure is not useful in disconnected networks, since there are nodes which are not connected, neither directly nor indirectly. Therefore, it is customary to compare only nodes that are in the same component, and not between nodes belonging to different components.

Let us examine Figure 8.6a. If we were to calculate the degree of the nodes, we would get the following result (Figure 9.5):

Thus, according to degree centrality, E is the central node, as it has the highest degree (4).

Table 9.1 A distance matrix

	A	B	C	D	E	F	G	H
A	0	1	2	3	4	4	5	5
B	1	0	1	2	3	3	4	4
C	2	1	0	1	2	2	3	3
D	3	2	1	0	1	1	2	2
E	4	3	2	1	0	1	1	1
F	4	3	2	1	1	0	1	2
G	5	4	3	2	1	1	0	1
H	5	4	3	2	1	2	1	0

Let us now calculate the closeness centrality of each of the nodes. To do this, follow these steps:

1) Calculate the distance between any two nodes.
2) For each node, sum its distances to all other nodes.
3) For each node, divide the number of nodes in the network (minus one) by the sum.

The result of the first step is presented in the following table, called a **distance matrix**:

Each cell contains the distance between the pair of nodes in the headers of its respective row and column.

The sum of each column is the sum of the distances between each node and all other nodes: A: 24, B: 18, D: 12, E: 13, F: 14, G: 17, H: 18.

The final step is to divide N-1 (in this case, 7) by the values we received.

The closeness centrality of each of the nodes will be:

$$A: \frac{7}{24} = 0.29, \ B: \frac{7}{18} = 0.39, \ C: \frac{7}{14} = 0.5, \ D: \frac{7}{12} = 0.58, \ E: \frac{7}{13} = 0.54,$$

$$F: \frac{7}{14} = 0.5, \ G: \frac{7}{17} = 0.41, \ H: \frac{7}{18} = 0.39$$

According to this measure, node D is the central node because it is relatively closer to all other nodes. The most marginal node is A, which is relatively far from all.

9.2.3 Betweenness centrality

Betweenness centrality measures the extent to which a node appears in the shortest paths between other pairs of nodes. This measure was first defined in 1977 by the sociologist Linton Freeman (1977). It is relevant in networks that involve the transport of information, news, messages, or rumors. The assumption underlying the measure is that traffic tends to move in the shortest path, and therefore, an object that is located within the path has great power. This can be likened to the strategic importance of seaports, straits, and canals, which constitute necessary passageways in the movement of goods. This was illustrated earlier with regards to the centrality of the Suez Canal.

The way to calculate this measure for a given node x is as follows:

- For each pair of nodes y and z that are not x, divide the number of short paths between y and z that include x by the total short paths between y and z.
- The sum of the results is the betweenness centrality measure of node x.

The final results can be very large, so it is customary to "normalize" them by dividing them by the number of pairs of nodes that are not x: $\dfrac{(N-1)(N-2)}{2}$. The normalized results are between 0 and 1, with a value of 0 reflecting a node that is not at the center of any path, and a value of 1 reflecting a node that is at the center of all paths.

We will demonstrate this on the network on which we modeled closeness centrality.

Node A is not found in any path, so its betweenness is 0. Recall that we are only considering the shortest paths.

Node B is only found in the paths that reach A, and in those it is found in all of them. Therefore, one must go through all the pairs of nodes that include A and do not include B and perform the calculation operation that we described earlier. Between A and C there is a single track, in which B is also found, so we will receive, according to step (1), the quotient $\dfrac{1}{1}$. Similarly, we get the quotient $\dfrac{1}{1}$ for the pairs A and D, A and E, A and F, and A and H. Conversely, between A and G there are two different paths, yet B is found in both. Consequently, the quotient will come out $\dfrac{2}{2}$. We will sum up the results and get the betweenness measure of B, $\dfrac{1}{1}+\dfrac{1}{1}+\dfrac{1}{1}+\dfrac{1}{1}+\dfrac{1}{1}+\dfrac{2}{2}=6$. A similar calculation will show that the betweenness measure of C is 10, and the betweenness measure of D is 12.

The cases of E, F, G, and H are different, because they are not located in necessary paths. We will demonstrate this with node E. It is positioned in the paths between each of the left nodes (A, B, C, and D) to nodes G and H and between nodes F and H.

Questions

5) *In which of the routes is it necessary to go through E if the goal is to take the shortest route?
   a. A to G
   b. B to G
   c. C to G

    d. D to G
    e. A to H
    f. B to H
    g. C to H
    h. D to H
    i. F to H

In the four cases where E is a necessary transition, the quotient will be 1. In the remaining five cases there is an alternate node to E: node F in paths a–d, and node G in path i. Consequently, half of the possible paths in these pairs pass through E, and thus the betweenness of node E is $6\frac{1}{2}$.

The betweenness value of each node is A:0, B:6, C:10, D:12, E:$6\frac{1}{2}$, F:2, G:$\frac{1}{2}$, H:0.

We will normalize the results by dividing the values by $\frac{(8-1)(8-2)}{2}$: A:0, B:$\frac{6}{21}$, C:$\frac{10}{21}$, D:$\frac{12}{21}$, E:$\frac{13}{42}$, F:$\frac{2}{21}$, G:$\frac{1}{42}$, H:0.

Let us compare the results we have received so far:

According to degree centrality, E is the central node with degree 4, and in second place there are three different nodes, D, F, and G. According to closeness centrality, D is more central, and E is in second place. According to betweenness centrality, node D is significantly more central (about 60% of all routes pass through it) and in second place is node C, despite its apparent marginality (based on other measures). The centrality of D and C stems from the fact that they are essential transition points on the way to A and B.

Henry Oldenburg (1619–1677) and Harriet Tubman (1822–1913) are two figures whose influence in their respective networks   – scientific and aboli-tionist – stemmed from their roles as brokers of information. Oldenburg, a theologian, scientist, and diplomat, moved across major European cities in the mid-17th

Table 9.2 Centrality results

	Degree centrality	Closeness centrality	Betweenness centrality
A	1	0.29	0
B	2	0.39	0.286
C	2	0.5	0.476
D	3	0.58	0.571
E	4	0.54	0.31
F	3	0.5	0.095
G	3	0.41	0.024
H	2	0.39	0

century, forging connections with scholars. In 1660, following the establishment of The Royal Society in London, one of the first scientific societies in Europe, Oldenburg was appointed secretary of the society. Oldenburg's role was to correspond with scientists across Europe and to serve as a conduit for the exchange of scientific information. His centrality in the Enlightenment, despite limited scientific contributions, arose from his role in facilitating communication between scientists across Europe, making him a key figure through high betweenness centrality in the network.

Similarly, Harriet Tubman, who was born around 1822 into slavery at Dorchester County, Maryland, became a leading figure in the anti-slavery movement as a bridge between runaway slaves from the South and free states in the North. Tubman escaped slavery in 1849, and settled in Philadelphia, Pennsylvania. However, she refused to simply enjoy her own freedom and became an active participant in the Underground Railroad. The Underground Railroad was a 19th-century informal network of abolitionists across the United States, who founded secret routes and safe houses that helped enslaved African Americans escape to free states or to Canada. It is estimated that Tubman personally guided approximately 70 individuals to safety. However, Tubman would not have achieved her success without the assistance of others that provided information and safety along her routes. Accordingly, her success as well as her centrality in the Underground Railroad network were based on her position as a broker. "It appears that her network position put her in an excellent position to dictate, control, and monitor the activities and information of the actors in the Underground Railroad network, thus reducing her 'structured ignorance'" (Young et al. 2009, 417). Much like Oldenburg, Tubman was in a position that gave her control over information.

9.2.4 Eigenvector measure

Although the degree of a node may be a good indication of centrality, its deficiency is in that it treats all connections as equal. As we have seen, a single edge to a central element in the system (closeness centrality) or a number of edges that make the node a necessary transition (betweenness) are more valuable than many edges at the margins of the network. If so, an alternative is a measure that prioritizes edges that connect to "important" nodes. Thus, according to this measure, a node is considered central if it is "prestigious", i.e., closer to other prestigious nodes.

In this section we will briefly describe two versions of the prestige measure: eigenvector centrality and PageRank centrality. Unlike the aforementioned measures, the calculation of prestige measures tends to be mathematically complicated, so we will not go into the details of the calculation.

The **eigenvector centrality** measure gives each node a score relative to the sum of its neighbors' scores, thus measuring the degree of influence of the node on the system. The first to use this measure was the German mathematician Edmund Georg Hermann Landau (1877–1938) in his 1895 article on the scoring method in chess competitions (Landau 1895).

PageRank is an algorithm developed for Google as part of the company's website ranking technology (Page et al. 1999). The company's founders, Larry Page and Sergey Brin, said that the algorithm was an intuitive attempt to describe the behavior of "random surfers": such surfers repeatedly click on links that take them from one occasional website to another. In other words, the PageRank index measures the chance that a random surfing of the web will lead to a given website.

Although this measure was created in response to the development of the internet, it is currently used to study social, historical, literary, geographical, biological, and other networks. For example, Matthew Jockers (2012) compared 3,592 novels published in Britain, Ireland, and the United States in the years 1780–1900. Using stylistic analysis tools, he identified typical discourse topics and created a network of similarity between the various novels. Jockers used the PageRank measure to identify the most influential authors in English literature, those for whom there is the highest chance that a "random reader" zapping between books will reach. According to him, Walter Scott (1771–1832) and Jane Austen (1775–1817) are the two most influential writers, or, in his own words, "the literary equivalent of Homo erectus, or, if you prefer, Adam and Eve".

9.2.5 Centrality measures and the Game of Thrones

We learnt about five centrality measures: degree, closeness, betweenness, eigen-vector, and PageRank.

Degree is a convenient and simple measure that summarizes the number of connections of each node. This measure is effective for cases where the quantity of ties is more important than their quality.

Closeness is a measure that refers to the "geographical" centrality of the node, how small the distance is between it and the others.

Betweenness is a measure that refers to the importance of the node as a bottleneck and as a strategic point in the system.

Eigenvector and PageRank are two measures that quantify the prestige of the various nodes by attributing different levels of importance to each and every node.

Let us demonstrate the difference by examining a network analysis of the first season of the Game of Thrones series, produced for HBO and based on the fantasy book series "A Song of Ice and Fire" by the writer George R. Martin (Figure 9.6). In this network created by Assaf Shapira (2020), each node represents a character in the series, and each edge represents a pair of characters who appear together in the same scene. A total of 86 nodes and 376 edges of different weights are included in the network.

In this network, each of the measures yielded a different center. The full results can be found at https://zefsegal.com/computational-literacy-for-the-humanities/. According to **degree**, the top four characters are Tyrion Lannister (34), Ned Stark (33), John Snow (25), and Kaitlyn Stark (25).

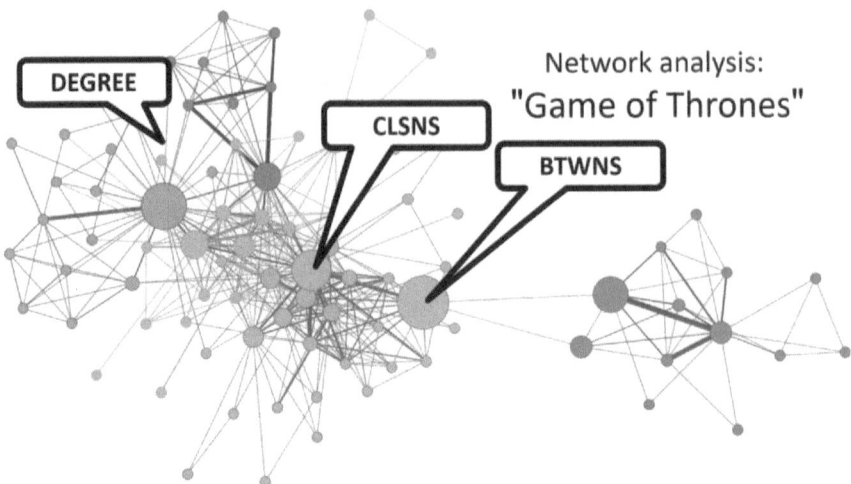

Figure 9.6 The *Game of Thrones* network.

Source: Shapira (2020).

Tyrion Lannister, the rejected and intelligent son of Tywin Lannister, is the main character in the series. The measure reflects the number of characters with which Tyrion Lannister appeared in scenes. Indeed, during the first season he travels around the kingdom and meets with a very wide range of characters, 34 to be exact. Along with him at the top is also Edward Stark, known as Ned, head of House Stark and Warden of the North, who met with 33 characters during the first season. In third and fourth place are two figures with a much lower degree: 25.

If we consider the weighted edge, that is the number of encounters of each character with other characters, and not just the number of characters they met, the main character, by far, is Ned Stark, who had 115 encounters with 33 of the characters in the series. Tyrion, on the other hand, had only 64 interactions, and thus drops to fourth place. John Snow is second with 84 interactions and Kaitlyn Stark third with 75.

Closeness reflects Ned's centrality in the plot. Although he finds his death at the end of this season, the plot of that season revolves around him. Moreover, the three main characters are the three characters fighting for control of the kingdom, Ned Stark (0.552), and the twin siblings Jaime (0.518) and Cersei (0.515) Lannister. Their closeness to the other characters, as we have argued before, reflects their importance to the whole plot.

Betweenness does not reveal the "heroes", but the brokers, those who are responsible for transmitting the information across the kingdom. Therefore, the betweenness centrality of Varys (0.263), the Master of Whisperers, should not surprise viewers of the series. Varys is responsible for collecting and storing information from unknown entities that are on the fringes of the Game of Thrones society, and therefore serves as a "necessary transition point". This is also evident from the graphic visualization of the network in Figure 9.6. His centrality as a mediator allows him to survive the upheavals of the series. All the other characters have substitutes, except him. In second and third places appear Tyrion Lannister (0.222) and Ned Stark (0.185) due to the large number of their interactions, but not necessarily because they are irreplaceable.

The two measures of **prestige** unequivocally determine the centrality of Ned Stark. According to the eigenvector measure, Ned Stark gets a value of 1, while in second and third places, the siblings Jaime and Cersei Lannister, are valued at 0.89 and 0.88, respectively. The large gap between first and second place is also present in the PageRank index, where Ned gets the value 0.055, while in second and third places, Daenerys Targaryen and Tyrion Lannister get only 0.044 and 0.041. Although the two measures identify the same central figure, each defines a different hierarchy. According to the eigenvector measure, the main figures are the heads of the royal family, and the people close to them. In contrast, according to the PageRank measure, the main characters are associated with three different plot lines, occurring in different geographic spaces: Ned Stark in the royal court, Daenerys Targaryen on the continent of Essos, and Tyrion Lannister across the continent of Westeros.

It is not possible to say that one measure is more correct than another, but only that they reflect different definitions of the question: What is the center?

9.2.6 Analyzing network centrality with Python

As we saw in the previous chapter, Python provides various tools to support network analysis (i.e., `degree()`, `shortest_path()`). In this section we will learn how we can use Python to apply various centrality measures to a network. However, before we begin, we will first examine the connectedness of the network. Measures such as closeness centrality are not useful in disconnected networks. If this is the case, we may choose to analyze each of the network's components individually, or possibly only its largest components.

Network connectedness

In the following code the method `is_connected()` is called with a network argument (G) and returns an appropriate Boolean value (True or False). If the network is not fully connected, the method `number_connected_components()` returns the number of components within the network.

```python
if nx.is_connected(G):
    print("The network is connected.")
else:
    print("The network has ", nx.number_connected_
    components(G), " components.")
```

The method `connected_components()` divides the graph into sets of nodes, one for each component. To find the largest component we call the function `max()` with two arguments: the divided graph (i.e., the output of `connected_components`) and the function that is used for comparison (`len` in our case).

```
components = nx.connected_components(G)
largest_component = max(components, key=len)
```

In order to analyze the network structure of one of the components we need to create a subgraph. In the following code we call the method `subgraph()` with the largest component of G as argument.

```
subG = G.subgraph(largest_component)
```

The variables G and subG store two networks: G is the entire network and subG is the largest connected component in G. Each of these variables can be used as arguments of functions and methods which require networks. In the following code, the two networks are printed on the screen by first calling the `draw()` method from the `networkx` package with either network as argument, and then calling the `show()` method from `matplotlib` to display the network. Note that in some environments, `draw()` automatically plots the network, making `show()` redundant.

```
import matplotlib.pyplot as plt
nx.draw(G, with_labels=True)
plt.show()

nx.draw(subG, with_labels=True)
plt.show()
```

Centrality measures

The `networkx` packages include methods for calculating centrality measures. The following three methods calculate closeness, betweenness, and eigenvector centrality respectively. They receive as an argument a network variable and return a dictionary in which the keys are the nodes, and the values are the centrality measure of each node.

- `nx.closeness_centrality()`
- `nx.betweenness_centrality()`
- `nx.eigenvector_centrality()`

As an example, consider the first four elements in the dictionary returned by `closeness_centrality()`, with the network in Figure 9.5, which was constructed in section 8.8, as input.

```
close_dict = nx.closeness_centrality(G)
```

```
{'A': 0.29, 'B': 0.39, 'C': 0.5, 'D': 0.58,...}
```

The key of each element is the node's label (A, B, C, etc. in Figure 9.5) and the value is its closeness centrality measure. The two other dictionaries have a similar structure, with the value representing other centrality measures.

In order to identify the five most central and the five most marginal nodes, we need to sort the elements according to the measure. However, because dictionaries cannot be sorted we need to convert the dictionary into a list of tuples, and use `sort()`. Since this method sorts the list according to the first coordinate of the tuple, when we convert the dictionary into a list of tuples we will switch the order. Recall that we followed a similar algorithm in section 5.3.2, when we created frequency lists of words within texts; we entered the "frequency" before the "word".

In the following code a loop iterates through the elements in `close_dict` and for each dictionary element it creates a new reversed tuple in `close_list`.

```
close_list = []
for k, v in close_dict.items():
    close_list.append((v, k))
```

The result is the following list:

```
[(0.29, 'A'), (0.39, 'B'), (0.5, 'C'), (0.58,
'D'),...]
```

This list can now be sorted in descending order, with the most central nodes appearing at the beginning of the list, and the most marginal one at the end.

```
close_list.sort(reverse=True)
```

Recall that the `sort()` method does not return a sorted copy of the list, but rather overrides the values of the list itself. Consequently, we can use indexing to retrieve different elements in the list according to their position. In the following code we print the three most central nodes.

```
for n in close_list[:3]:
    print(n)
```

9.2.7 Centrality in the E-road network

This exercise picks up where we left off in section 9.1.2. We used Python to create the network of European E-roads, and stored it in the variable G. In this exercise, we will write code that examines the connectedness of the network and that identifies the most central cities in the network according to four centrality measures (degree, closeness, betweenness, and eigenvector) and then interpret our results.

Questions

6) Connectedness and connected components:
   a. *Print the number of components in the E-road network.
   b. *Iterate through the components and for each component print the number of nodes using `number_of_nodes()` and display its network visualization on the screen.
   c. How many components is the road network divided into?
   d. What characterizes the three largest components?

The largest component includes the European continent and a significant part of the Asian continent. The second and third largest components (Figure 9.7) are the islands of Great Britain and Ireland.

Questions

7) Degree centrality:
   a. *Write a program that prints the ten most central nodes in the E-road network in terms of degree centrality. Hint: Create a list of tuples (degree, node). Use the `degree()` method to calculate for each node its degree. Recall that `G.degree("Paris")` returns the number of E-roads passing through Paris.
   b. What characterizes the most central nodes in the network in terms of degree?

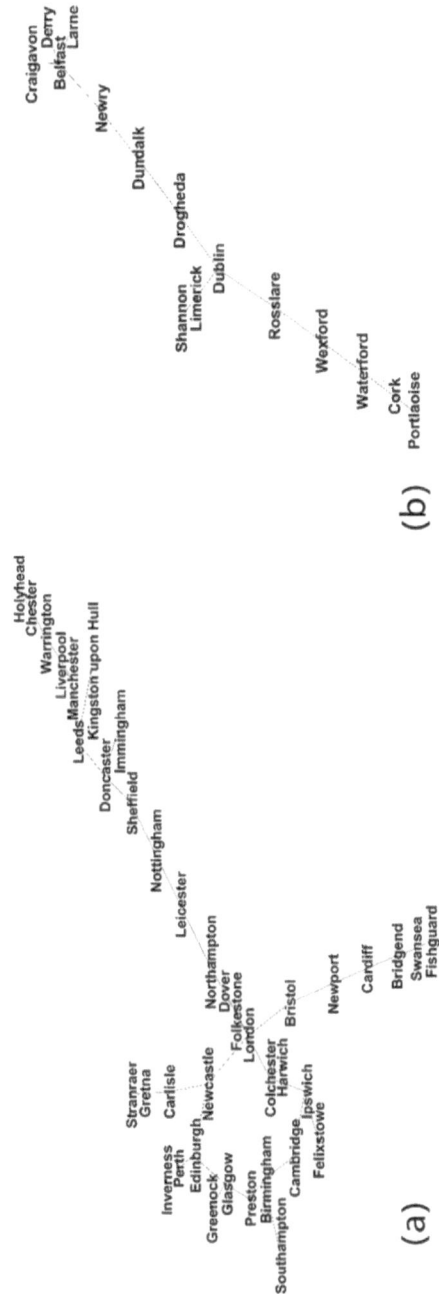

Figure 9.7 The second (a) and third (b) largest connected components of the E-road network.

The ten most connected cities in the E-roads network are Moscow, Berlin, Paris Vienna, Munich, Prague, Liege, Metz, Budapest, and Bratislava. With the exception of Munich, Liege, and Metz, all of these cities are capital cities, which could be a simple explanation for their degree centrality.

Questions

8) *Create a subgraph of the largest component in the E-road network.
9) Closeness centrality
   a. *Identify the five most central cities and the five least central cities in the largest component of the E-roads network in terms of closeness centrality.
   b. What characterizes the cities in each category?

The five most central cities in the largest component of the E-roads network in terms of closeness centrality are Warsaw, Lviv, Kaunas, Lublin, and Poznan. The five least central cities are Helleland, Sandnes, Stavanger, Randaberg, and Rennesøy.

The most central cities are located in Eastern Europe (in the region of Poland, Lithuania, and Ukraine), while the least central cities are located in Norway.

Questions

10) Betweenness centrality:
    a. *Identify the five most central cities and the five least central cities in the largest component of the E-roads network in terms of betweenness centrality.
    b. What characterizes the cities in each category?

The five most central cities in terms of betweenness in the main component of the network are Warsaw, Saint Petersburg, Kaunas, Berlin, and Poznan. The five least central are Astara, Antirrio, Antalya, Altenmarkt im Pongau, and Aktio.

When examined closely, we see that the most central cities are located on an interesting path. Except for St. Petersburg, which is a transition point to Scandinavia, the remaining four cities are located on one axis that crosses Poland, and thus form a major connection between Western Europe and the Far East.

The most marginal cities are located at the edges of highways: Astara in the South-East corner of Azarbeijan, Antirio in South Greece, Aktio in West Greece,

Antalya in South Turkey, and Altenmarkt im Pongau in central Austria. What is more striking is the fact that they are organized alphabetically. Since we only have the letter "A", we can assume that there are many more marginal cities, besides for these five. Every highway has an end, and that end would be an additional marginal city that is not in-between any other two cities.

Questions

11) Eigenvector centrality
   a. *Identify the five most central cities and the five least central cities in the largest component of the E-roads network in terms of eigenvector centrality.
   b. What is the difference between the cities that were found to be central according to the previous measures and between the main cities according to the eigenvector measure?

The five cities with the highest eigenvector measure are Paris, Metz, Reims, Brussels, and Liege. The five with the lowest measure are Helleland, Sandnes, Stavanger, Randaberg, and Rennesøy.

In contrast to the central cities we identified previously, the eigenvector measure emphasizes a very limited area – Belgium and northern France. The uniqueness of this area is reflected by a large density of E-roads, which can also be clearly seen in the map of the road system. To a large extent, this measure identifies density of connection within the European road system.

9.3 Groups of nodes

Analyzing the structure of a network does not only provide information about individual nodes, but also yields information regarding communities – relatively cohesive groups within a large and decentralized system. An example of this is social networks, which tend to create groups whose members have stronger and more frequent connections with members of those groups than with those who do not belong to them. In many networks there is a natural division into groups or communities. Thus, for example, the internet can be divided according to groups of densely linked sites, and social networks can be divided according to groups of friends or acquaintances. In most cases, however, the identification is not so simple and requires an in-depth computational analysis of the various relationships. The way the network splits into communities may reveal its structure and governing rules, which would be difficult to identify otherwise. This is why much of the research in the field of networks deals with questions regarding the characterization and identification of sub-communities.

9.3.1 "Natural" divisions – connected components and cliques

The divisions we will now define are natural, because the connection between the members of each group is so clear that it does not need further explanation. The most trivial division is that of connected components (see section 8.6). Due to the fact that between any two nodes in the same component there is a path and that between any two nodes that are not in the same component there is no path, the split is warranted. On the basis of such a division it was argued in the beginning of Chapter 8 that an analysis of the network of kinship and genetic ties between rulers in 1618 reveals the rival sides in the Thirty Years' War. Similarly, the division into connected components of the E-road system creates a "natural" distinction between landmasses, and in particular separates the British island from the continental landmass.

The Hungarian writer Frigyes Karinthy claimed in his short story "Chains" (Láncszemek), from 1929, that between every two people on earth there is a chain of at most five acquaintances (Karinthy 1929). In 1967, the American psychologist Stanley Milgram established this claim scientifically in a study of the population of the United States (Milgram 1967). The common term "six degrees of separation" was coined by playwright John Guare in 1990. Common to all of these is the claim that the inhabitants of earth are part of a connected network, where the distance between nodes is not particularly large. Nevertheless, it is clear to us, from our acquaintance with human society, that there are more cohesive groups within the connected system.

A clique is a group of nodes in which every node is connected to all others with an edge. Unlike connected components, there is no restriction against clique members' having additional neighbors who do not belong to this cohort, or even belonging to other cliques. In Figure 9.8, for example, there is a clique of four nodes, with another node adjacent to one of the members.

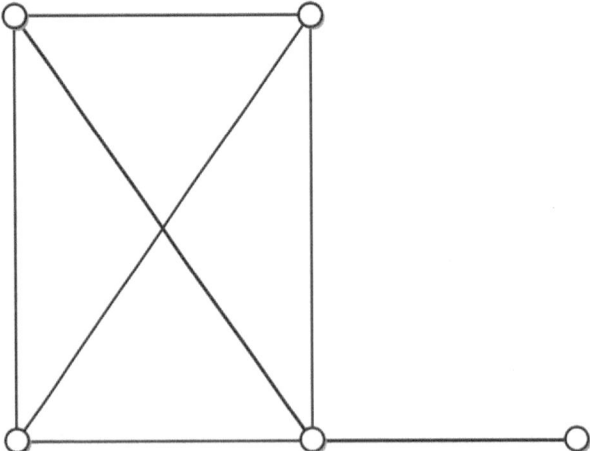

Figure 9.8 A network with a clique.

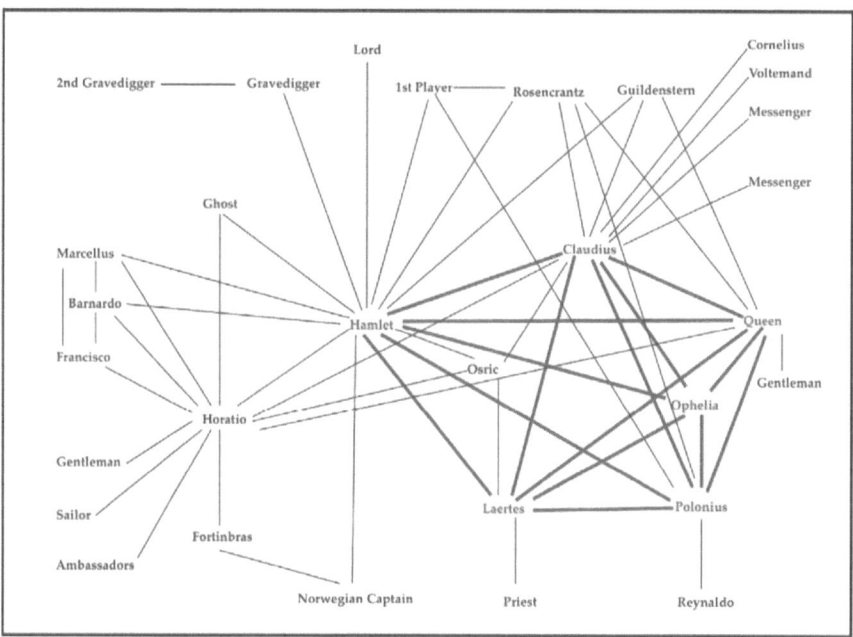

Figure 9.9 The Hamlet network.

Source: Moretti (2011).

Many times, the existence of a clique within a sparse network is evidence of the existence of a cohesive subgroup, whose power and influence will stem from this formation. For example, in a network representing the interactions in the play Hamlet there is a clique, whose edges are emphasized in Figure 9.9 (Moretti 2011). The clique includes the characters Hamlet, Claudius, the Queen, Ophelia, Laertes, and Polonius – this is the royal court of the Danish monarchy in the fictional world created by Shakespeare; this is the elite of this society.

9.3.2 Modularity

For the most part, relationships between community members are not close enough to be considered a clique, yet they are closer than their relationships with elements outside the community. Community identification techniques, called **modular decomposition,** rely on a variety of methods that identify "close" or "dense" areas of the network. These algorithms work by breaking down the network into tighter communities. Mathematically, the success of modular decomposition is reflected in the fact that the density of the connection within the sub-network (modular unit) is higher than the density of the connections between the sub-networks. Nevertheless, this success does not help to interpret the significance of the decomposition. As

Figure 9.10 A modular decomposition of the E-road network.

always, the meaning of this decomposition must come from a deeper interpretation of the properties of the nodes and the network.

Modular decomposition is useful because it quickly divides complex networks into subnetworks, which are easier to analyze than the hundreds, thousands, or millions of nodes and edges in the original network. Due to the complexity of the modular decomposition process, we will not present here the algorithm but only the results. It is important to clarify that the various modular decomposition algorithms rely on statistical models, and therefore the results are not unambiguous. Each run of such an algorithm will result in slightly different modular units, especially at the margins of these units.

We will demonstrate modular decomposition on the E-road system. Running a modular algorithm on the network using Gephi software identifies 47 different communities. In Figure 9.10, the shades of the nodes representing the cities reflect the modular units to which they belong. As can be seen, the division represents quite faithfully the geographical location of the cities and the spatial structure of European society.

In the modular decomposition of the *Game of Thrones* network the 86 characters are divided into communities: 8 characters belong to the Night's Watch located on the northern border, 27 characters in the royal court, 20 characters in the northern province, 13 characters on the Asus continent, and 16 characters belonging to other districts in the kingdom. (The results can be found at https://zefsegal.com/comput ational-literacy-for-the-humanities/.) The resulting modular decomposition is not the analysis and interpretation, rather, it is a basis for such an analysis. In this case, we used modularity as a basis for spatial, social, and class analysis of society in a television series.

9.4 Network patterns

9.4.1 Density and the clustering coefficient

Often, identifying nodes that belong to one clique, or a number of cliques is more important than identifying the cliques themselves. On the one hand, a node belonging to a number of cliques has an influence on strong communities, but on the other hand, it is substitutable, and therefore its effect is smaller. In this section we will introduce two indicators that make it possible to quantify the density of a network and the density of the surroundings of a node, in order to evaluate the influence of individual nodes and cliques on the system.

Density is defined as the ratio between the number of actual edges in the network and the number of possible edges. The formula for the density of a network with N nodes is $\dfrac{|E|}{\left(\dfrac{N(N-1)}{2}\right)}$, with E being the set of edges. For example, the network depicted in Figure 9.8 has 8 out of $\dfrac{5(5-1)}{2}$ possible edges; the density is therefore $\dfrac{8}{10} = 0.8$.

In comparison, the density of the E-road network is much lower: 0.002. The reason is that the highway system is not designed to cover all possible connections.

For this purpose, there are smaller roads that connect the different cities.

Daniil Skorinkin, in a 2017 article, used the density measure to analyze Lev Tolstoy's book *War and Peace* (Skorinkin 2017). Skorinkin's network connects two characters whenever they appear together in the same sentence. His findings show that the parts dealing with war are characterized by a relatively sparse network of characters (0.15 on average), while the parts dealing with peace are characterized by a relatively dense network (0.25 on average). The correlation coefficient between the density of the network in a particular chapter and the extent to which it is a chapter on peace or war is approximately 0.65. Recall that this is a fairly high correlation (see Chapter 7). Skorinkin speculates that the reason for this correlation is that war descriptions tend to include fewer direct interactions between characters, and thus the networks of these parts are sparser. This claim is consistent with other studies that have shown that tragedies tend to create sparser networks compared to comedies (Trilcke et al. 2015).

The concept of density can also be used to broadly characterize the surroundings of given nodes. We can ask to what extent do the neighbors of a given node tend to be n eighbors of each other. This measure is called the **clustering coefficient.** The formula for calculating this coefficient for node x is: $\dfrac{\textit{The number of pairs of neighbors of x that are neighbors of each other}}{\textit{The number of pairs of neighbors of x}}$. That is, one must go through all the pairs of x's neighbors and count how many of them are neighbors themselves.

In a 2017 paper, Cornell Jackson analyzed the structure of the Scottish elite in the years 1093–1286 by using networks that described the ties between elite members, with an edge appearing between two members if both co-signed an alliance or a contract (Cornell 2017). According to Jackson, a low clustering coefficient indicates the importance of a member in the Scottish elite, due to his vital role in connecting two other elite members. The clustering coefficient makes it possible to prove the importance of Donnchad IV, Earl of Fife (1113–1154) and Walter fitz Alan (1106–1177), two nobles who worked under the patronage of David I (Dauíd mac Maíl Choluim), king of Scotland (1084–1153).

9.4.2 Degree distribution

The distribution of the degrees of nodes over the whole network is another way to understand the structure of a network. The distribution combines a close look at the nodes themselves (their degrees) and a distant view of the entire network (the distribution of all the degrees). For example, the degrees in the network depicted in Figure 9.5 are (1,2,2,2,3,3,3,4). Thus, the value 1 occurs one-eighth of the time, the value 2 occurs three-eighths of the time, the value 3 occurs three-eighths of the time, and the value 4 occurs one-eighth of the time. This can be plotted as a scatter plot showing the relative share of each value, or as a bar chart showing the number of returns of a particular value. (See Figure 9.11.)

Like the other measures presented so far, the distribution of degrees, by itself, is not sufficient to understand the structure of the network, but it sometimes reveals phenomena that are not apparent at first glance.

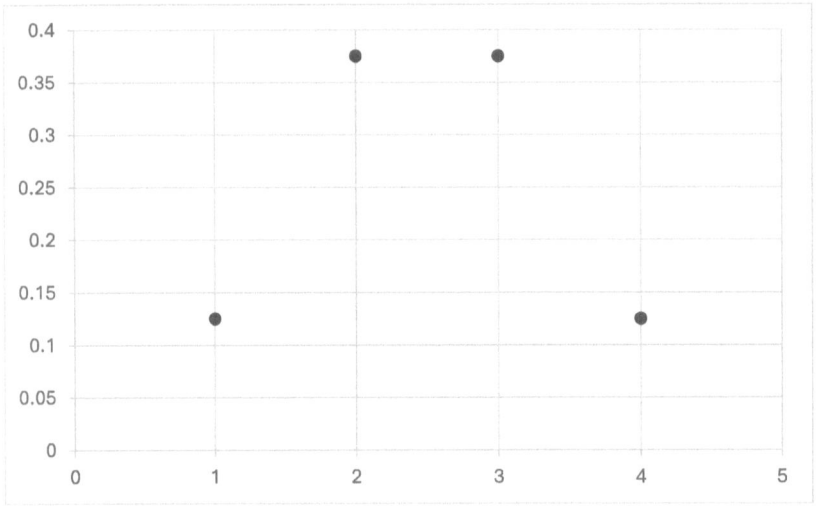

Figure 9.11 Degree distribution (scatter plot).

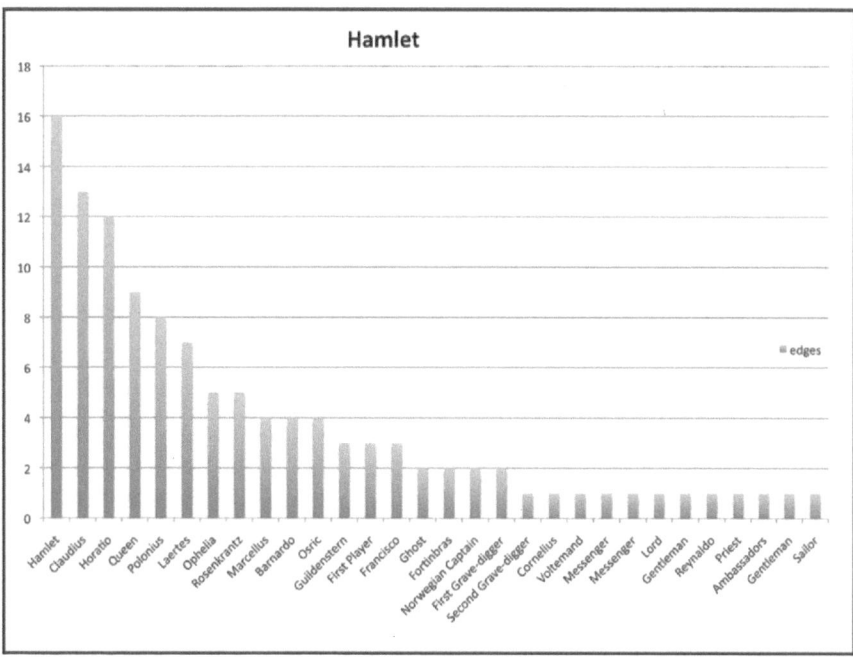

Figure 9.12 The degree distribution of the character network in *Hamlet*.

In 1999, Albert-László Barabási proved that the distribution of degrees in real-world networks is not usually uniform but rather unequal and "long-tailed". In these distributions, only few nodes have high degrees, whereas the rest have low degrees. These distributions are called scale-free distributions if the resulting diagram has an exponential power-law curve. We have already seen this phenomenon in Chapter 5 when we learned about Zipf's Law.

For instance, Figure 9.12, which depicts the degree distribution of Hamlet's social network, shows that Shakespearean tragedies follow a pattern similar to word frequency in language. This is not a coincidence and holds true for many networks. A 2002 study of the network of superheroes from the Marvel universe (6,486 characters and 168,267 edges) showed that the world of comics operates according to the same network rules, even though it was created almost randomly over a 40-year period by a group of creators with very little premeditation and intent (Alberich, Miro-Julia, and Rosselló 2002). With the exception of a small number of characters, such as Captain America and Spider-Man, whose degree is about 2,000, the vast majority of the characters have few interactions with others.

9.4.3 Analyzing network patterns with Python

Modular decomposition

There are a number of algorithms for modular decompositions of graphs, we will use the method `best_partition()` that accepts a network parameter and a value defining the required resolution and returns a modular decomposition. The resolution changes the size of the communities in the resulting decomposition. As previously mentioned, since the algorithm relies on a statistical model, the outcome might change slightly every iteration.

The method is included as part of the `community_louvain` package, so it needs to be imported.

```
import community.community_louvain as cm
partition = cm.best_partition(G, resolution = 0.6)
```

The output is a variable `partition` with a structure of a dictionary. The keys are the nodes, and the value of each key is a number identifying the modular unit (the unit identifier). Thus, for example, Glasgow and Birmingham belong to the same community, number 1, and Paris and Bordeaux belong to a different community, number 3.

```
partition:
{'Glasgow': 1, 'Birmingham': 1, 'Paris': 3, 'Madrid': 4,
'Southampton': 1, 'Le Havre': 3, 'Bordeaux': 3,
'Seville': 4,...}
```

In order to explore the output and check whether the modular units could be interpreted we will convert it to a more convenient format. Instead of the dictionary `partition`, in which the key is the node, and the value is the unit identifier, we will create a new dictionary – `communities` – in which the key is the unit identifier, and the value is a list of nodes belonging to the corresponding unit. We will write the code for this procedure in the next section (9.4.4).

```
communities:
{1:   ['Glasgow',   'Birmingham',   'Southampton',
'Le   Havre',...],   4:   ['Madrid',   'Seville',...],
3: ['Paris', 'Bordeaux',...],...}
```

Density

An additional property of networks is their density. The method `density()`, which is part of the `networkx` package, takes as an argument a graph value and returns the density measure of the graph.

```
import networkx as nx
nx.density(G)
```

The density of the E-road network is 0.002.

Degree distribution

Degree distribution is another network property that we can examine using Python. For this, we first need to create a list of the degrees of all the nodes in the network.

The following code iterates through the nodes of the network and appends the degree of each one to a new list variable – degrees.

```
degrees = []
for n in G.nodes():
    degrees.append(G.degree(n))
```

```
degrees:
[1, 3, 2, 3, 1, 2, 8, 5, 4, 3, 5, 2, 6, 3, 2, 2, 4,
2, 3, 2, 2, 2,...]
```

We will use the method `hist()` included in the library `matplotlib.pyplot` to draw a histogram depicting the degree distribution. The X-axis will represent the degrees and the Y-axis their frequency (or, in other words, the relative share of each degree value).

```
import matplotlib.pyplot as plt
plt.hist(degrees)
plt.show()
```

The histogram in Figure 9.13 represents the degree distribution of the cities in the E-road network and reflects Barbasi's claim regarding the shape of real-life

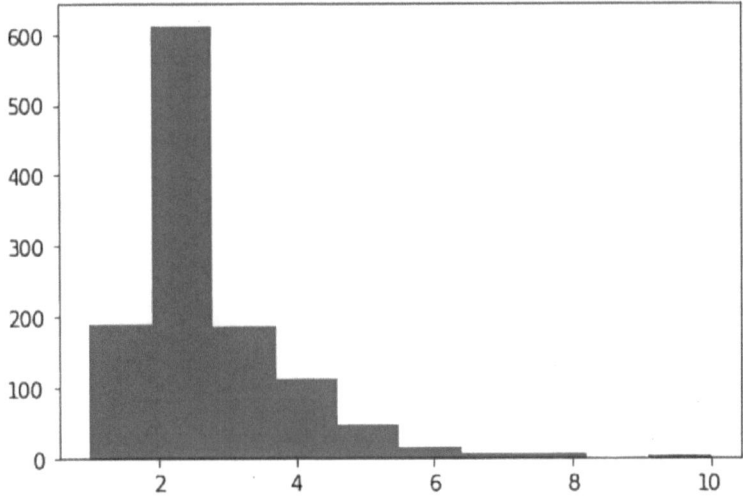

Figure 9.13 The degree distribution of the cities in the E-road network.

networks. According to the histogram, most cities (ca. 600) have degree 2 and are thus on a path connecting two other cities. Approximately 200 cities are located at the edge of the highways and therefore have degree 1. On the other hand, very few cities are adjacent to more than 6 other cities. These are the central hubs of the highway system.

9.4.4 Patterns in the Eurovision Song Contest network

The Eurovision Song Contest is an annual competition held since 1956 at the initiative of the European Broadcasting Union. The participants in the contest represent the member countries of the Broadcasting Union, and they are ranked at the end of the competition according to the score they receive from each of the other countries.

Although the Eurovision Song Contest was founded with the goal to unite Europe and transcend political differences, there are many who argue that the opposite is true. In an article published on May 15, 2019 on ABC Australia about Israel's hosting the Eurovision Song Contest, the headline "Eurovision was meant to be apolitical. It's been anything but" summarized the view according to which the competition has been political since its inception (Watson and Slee 2019). These politics are also the subject of the book *A Song for Europe: Popular Music and Politics in the Eurovision Song Contest* (Raykoff and Tobin 2007). As the editors of the book claim in the introduction: "From its inception, the Eurovision has seemed to reflect the political zeitgeist of Europe, even to anticipate certain political developments" (Raykoff and Tobin 2007, 1). In the following exercise, we will seek to empirically examine the claim of the authors of the article on the

ABC Australia network and the editors of the book, and answer the question: Are Eurovision voting patterns politically or geographically biased?

The data

To answer this question, we have created a weighted network in which the nodes represent the different countries that have participated in the Eurovision Song Contest over the years, and the weight of the edge connecting two countries is the average score that the countries have given each other. This information is stored in a CSV (comma-separated values) file that can be found at https://zefsegal.com/computational-literacy-for-the-humanities/.

The CSV file contains three columns. The first two columns are names of countries, and the third column is the average score these countries gave each other. Each row defines an edge: the two nodes that it connects and its weight. This information is enough for Python to automatically create the node list.

Building a network with weighted edges

The algorithm for building this network is similar to the one we used to build the E-road network with one caveat – the edges in the Eurovision network have weights.

Python reads CSV content as strings and each row that is read is stored as a list of three strings even though the third element, the weight of the node, is numerical.

```
['Armenia', 'Azerbijan', '0.06122449']
```

Consequently, we cannot simply convert each row to a tuple and append it using `edges.append()`, which was used for building the E-road network. We will need to convert the third string into its corresponding numerical value and create a tuple of the following form:

```
('Armenia', 'Azerbijan', 0.06122449)
```

To do so, we will use the function `float()`, which receives a string and converts it to a decimal number (contrary to the function `int()` that converts strings into whole numbers). The following loop parses each row to 3 separate values: two strings, for the node labels, and one numerical value for the weight.

```
for row in list(csv_content)[1:]:
    edges.append((row[0], row[1], float(row[2])))
edgecsv.close()
```

Dividing the network to communities

As described in the previous section, best_partition() can be used to perform modular decomposition of a network. However, in order to analyze the network, it is recommended to convert the output of this function (partition, below) to a more convenient format (communities, below), in which in each element the key corresponds to a community and the value, to a list of nodes which belong to the community.

```
partition:

{'Armenia': 6, 'Azerbaijan': 6, 'Andorra': 2, 'Bosnia
& Herzegovina': 1, 'France': 3, 'Slovakia': 6,...}

communities:

{1: ['Bosnia & Herzegovina',...], 2: ['Andorra',...],
3: ['France',...], 6: ['Armenia', 'Azerbaijan',
'Slovakia',...],...}
```

This restructuring of the data can be executed according to the following algorithm:

Define a new dictionary variable: communities
Iterate through all the items in partition and do the following:
    Store the value of the item (the unit identifier) in a variable unit.
    [1] Check if communities has an item whose key is equal to unit (e.g., 3),
        If the answer is yes,
            [2] append current node (e.g., "Bordeaux") to this item list value
        If the answer is no,
            [3] add a new item to communities:
                the key is unit (e.g., 3)
                the value is a list with node (e.g., ['Paris'])

The implementation of this algorithm requires the use of 3 dictionary methods:

[1] Retrieve the value of a specified key (if exists)

```
unit = partition.get(node)
```

[2] Append an item to the list value of item with the specified key

```
communities[unit].append(node)
```

[3] Add a new item to the dictionary with specified key and list with one item

```
communities[unit] = [node]
```

Questions

12) The Eurovision network:
   a. *Create a network based on the data in "eurovision_edges.csv"
   b. *Divide the network into communities (try different resolutions between 0.6 and 0.7).
   c. *Print each community in a separate line.
13) Examine the resultant communities.
   a. Can they be characterized as political or geographical?
   b. Are there any exceptional countries in the community?
   c. If so, is there an explanation for the exceptionality?

9.5 Building lexical networks with Python

In this chapter we considered networks which represent the relationship between cities (the E-road network and the Inca road network), train stations (Hankyu railway lines), countries (the Eurovision network), and literary characters (*Game of Thrones*, *Hamlet*, and *Macbeth*). Our final network is a lexical network that represents relationships between words.

The nodes in this network represent the nouns in a text, and the edges connect two nouns if they appear together in the same sentence. The weight of the edges is the number of joint occurrences of the two nouns throughout the text. We have covered most of the required tools and techniques in this and in the previous chapter.

9.5.1 Lemmatization

One Natural Language Processing concept that has not yet been introduced is **lemmatization**. For the purpose of creating a noun network will use a POS tagger to single out the nouns from other words in the sentence (see also section 7.2.2). In addition, in order to optimize the informativeness of the network, we will execute an additional natural language processing step called lemmatization. With lemmatization, one identifies the base form (lemma) for each word. For example, the base form of nouns in the plural (e.g., 'cats') is the noun in the singular ('cat'). The base form of a conjugated verb in English (e.g., 'chased') is the infinitive form ('chase'). Recall that we performed a similar process (manually) in section 2.2.2, when we defined a method for creating a concordance.

The `lemmatize()` method receives as input a word form and its POS and returns its respective lemma. This method is part of the WordNetLemmatizer package (see next page).

```
wnl.lemmatize("cats", "n")
wnl.lemmatize("chased", "v")
```

Consequently, in order to lemmatize words, we first need to identify their POS. We will focus here only on nouns ("n") and verbs ("v") and use the Penn Treebank tagset, which we used in section 7.2.2 to label different types of nouns and verbs.

- **NN**: Noun, singular or mass (e.g., 'cat')
- **NNS**: Noun, plural (e.g., 'cats')
- **NNP**: Proper noun, singular (e.g., 'James')
- **NNPS**: Proper noun, plural (e.g., 'Vikings')
- **PRP**: Personal pronoun (e.g., 'she')
- **PRP$**: Possessive pronoun (e.g., 'his')
- **VB**: Verb, base form (e.g., 'get')
- **VBD**: Verb, past tense (e.g., 'got')
- **VBG**: Verb, gerund or present participle (e.g., 'getting')
- **VBN**: Verb, past participle (e.g., 'gotten')
- **VBP**: Verb, non-3rd person singular present (e.g., 'get')
- **VBZ**: Verb, 3rd person singular present (e.g., 'gets')

Lemmatization – step by step

We begin the lemmatization process by importing the `nltk` library, from which we will download the relevant tools.

```
import nltk
```

We then download the `punkt` module and tokenize our text by calling the `word_tokenize` function.

```
nltk.download("punkt")
tokenized = nltk.word_tokenize("My dog chased the cats.")
```

```
tokenized:
['My', 'dog', 'chased', 'the', 'cats', '.']
```

Next, POS-tagging requires us to download the `averaged_perceptron_tagger` and `wordnet`, then call `pos_tag()` with the tokenized text as its argument. The function returns a list of tuples, where each tuple consists of a word and its POS tag.

```
nltk.download("averaged_perceptron_tagger")
nltk.download("wordnet")
tagged = nltk.pos_tag(tokenized)
```

```
tagged:
[('My', 'PRP$'), ('dog', 'NN'), ('chased', 'VBD'),
('the', 'DT'), ('cats', 'NNS'), ('.', '.')]
```

Finally, we import and define the lemmatization tools from the `nltk` library. Then, our code loops through the tuples in `tagged` and creates a list of corresponding `lemmas`. For each tuple it examines whether the first character in the tag is "N" (for all types of nouns) or "V" (for all types of verbs). If it is, it calls the `lemmatize()` method with the word and the lowercase form of the first character. Otherwise, it appends the word itself to the `lemmas` list.

```
from nltk.stem import WordNetLemmatizer
wnl = WordNetLemmatizer()

lemmas = []
for word,tag in tagged:
    if tag[0] == "N" or tag[0] == "V":
        pos = tag[0].lower()
        lemmas.append(wnl.lemmatize(word, pos))
    else:
        lemmas.append(word)
```

The result is a list of the lemmas of the words in our example sentence.

```
lemmas:
['My', 'dog', 'chase', 'the', 'cat', '.']
```

9.5.2 Creating a noun co-occurrence network

We will divide the process into four steps:

1. Extracting nouns
2. Calculating co-occurrences of nouns in sentences
3. Creating a list of edges
4. Creating the network

We will demonstrate the process on a simple text containing three short sentences.

Step 1: Extracting nouns

The input is the text.

```
text:
My dog chased the cats. One cat escaped and hid in
a box. Cats like to hide in boxes.
```

The output is nountext: a version of the text in which each original sentence is represented as a list of its noun lemmas.

```
nountext:
[['dog', 'cat'], ['cat', 'box'], ['cat', 'box']]
```

The algorithm (Figure 9.14):

Begin with the text stored as a string
Call sent_tokenize(text) and get a list of sentences.
Loop through the list of sentences. For each sentence (sent):
      Call word_tokenize(sent) and get a tokens list
      Call pos_tag(tokens) and get a tagged list of tuples

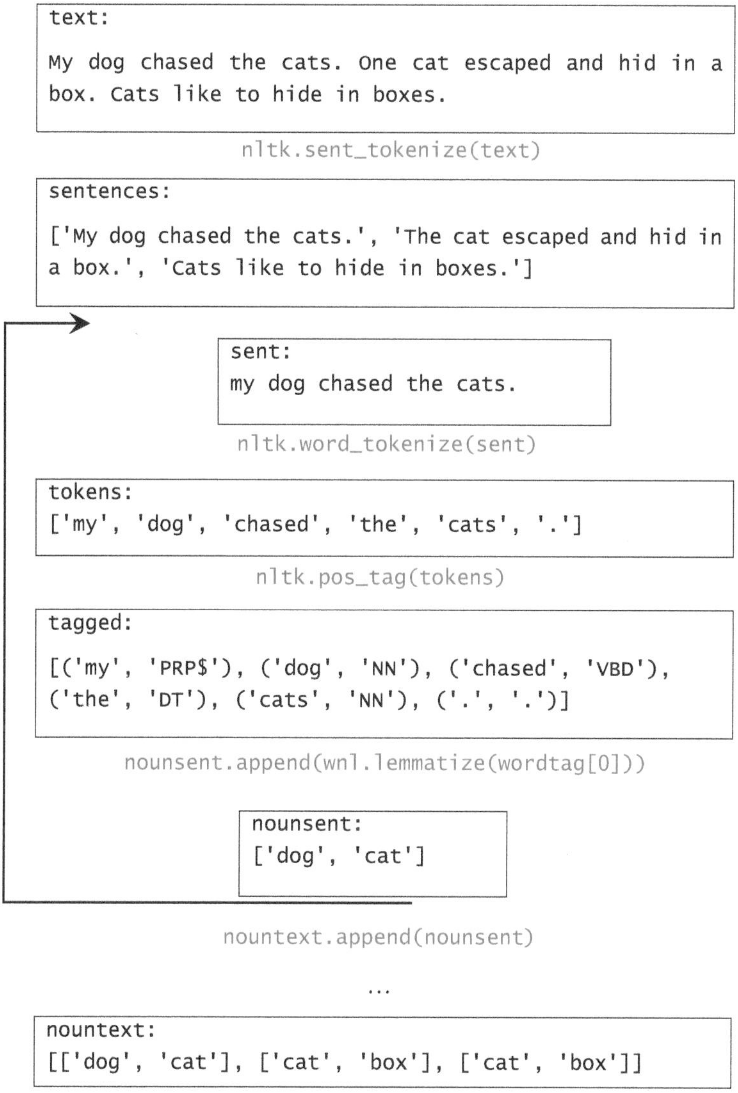

```
text:

My dog chased the cats. One cat escaped and hid in a
box. Cats like to hide in boxes.
```

nltk.sent_tokenize(text)

```
sentences:

['My dog chased the cats.', 'The cat escaped and hid in
a box.', 'Cats like to hide in boxes.']
```

```
sent:
my dog chased the cats.
```

nltk.word_tokenize(sent)

```
tokens:
['my', 'dog', 'chased', 'the', 'cats', '.']
```

nltk.pos_tag(tokens)

```
tagged:

[('my', 'PRP$'), ('dog', 'NN'), ('chased', 'VBD'),
('the', 'DT'), ('cats', 'NN'), ('.', '.')]
```

nounsent.append(wnl.lemmatize(wordtag[0]))

```
nounsent:
['dog', 'cat']
```

nountext.append(nounsent)

...

```
nountext:
[['dog', 'cat'], ['cat', 'box'], ['cat', 'box']]
```

Figure 9.14 From text to lists of noun co-occurrences.

Iterate through the tagged list. For each tuple (`wordtag`)
    if the second item in the tuple begins with "N"
        lemmatize the first item and append `nounsent` list
Append `nounsent` to `nountext` (a list where each item is a list
of nouns which correspond to one sentence in the text).

Step 2: Calculating co-occurrences of nouns in sentences

The input to this step is the output of the previous step (`nountext`).

```
nountext:
[['dog', 'cat'], ['cat', 'box'], ['cat', 'box']]
```

The output is d, a dictionary whose keys are pairs of noun lemmas which appear together in sentences, and the value is the number of their co-occurrences. For example, the lemmas 'box' and 'cat' appear twice in the same sentence in our text.

```
d:
{('cat', 'dog'): 1, ('box', 'cat'): 2}
```

The algorithm:

Create an empty dictionary d
Loop through `nountext` (sentences with only nouns). For each sentence
(`sent`):
    Iterate through all the indexes of the words in the sentence. For each
    index (`i`):
        Store the word in the `i` position in the variable `token`
        Iterate over all the following words (from position `i + 1`
        to the end of the sentence). For each word `t`:
            Define a key – the tuple (`token, t`)
            If this key already exists in d
                Increase its value by 1
            Else
                Create a new item with this key and value 1

Step 3: Creating a list of edges

The input to this step is the dictionary that we created in the previous step.

```
d:
{('cat', 'dog'): 1, ('box', 'cat'): 2}
```

The output is edges, a list of tuples which represent edges in the noun network: the two nodes (i.e., nouns that co-occur) and the weight of the edge between them (i.e., the number joint occurrences).

```
edges:
[('cat', 'dog', 1), ('box', 'cat', 2)]
```

The algorithm:

```
Create an empty list edges
Iterate through dictionary d. For each item:
        Create a triplet (noun, noun, weight)
        Append it to edges
```

Step 4: Creating the network

The input is a list of edges.

The output is a network in which the nodes represent noun lemmas in the text, and the edges between two nodes represent the number of times the two nouns occur in the same sentence.

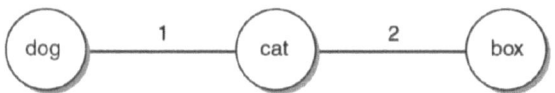

9.6 Distant reading: Relationship between nouns in Wharton's and Stein's texts

We will conclude this chapter with an exercise that constitutes the last tier of the distant reading of the four literary works authored by Edith Wharton and Gertrude Stein that we have been performing since Chapter 6. In this exercise, we will create a network that represents the relationship between nouns in each text. The nodes in the network will represent the nouns in the text, and the edges will connect two nouns if they appear together in the same sentence. The weight of the edges will be the number of joint occurrences of the two nouns throughout the text.

Questions

14) Building the noun network:
   a. *Before you begin, you must "clean" the text:
      • The end of a line in a text file might be read as the end of a sentence. In order to avoid this confusion, replace new line characters ("\n") with spaces (" ").
      • Wharton's novel has a number of sequences of dots with spaces between them. The tokenizer divides these sequences into empty sentences. Erase them by replacing " ." with ".".
   b. *Following the algorithms described in section 9.5.2. write a program that builds noun networks for the four literary works.
15) Analyzing the networks:
   a. *Update the code to sort the edges by their weight and print the top ten edges.
   b. *Update the code to identify and print the top ten nodes ranked by degree centrality.
   c. *Check if the graph is fully connected. If not, calculate the number of components, and the number of nodes and edges in the largest component.
   d. *Implement code to compute and print the top ten nodes based on closeness centrality.
   e. *Determine the density measure of the network.
   f. *Create a histogram representing the degree distribution of the nodes in the network.
16) Comparing the networks:
   a. Why is the degree distribution similar in each of the four texts?
   b. Why is the density of all four networks very low, and what does the relatively high density in the network of Stein's novel tell us?
   c. Look at the key nouns according to degree and closeness centralities, and answer the following questions:
      i. Is it possible to identify a difference between the key nouns in the two autobiographies and the nouns in the two novels?
      ii. What characterizes the key nouns in Edith Wharton's novel?

The four networks represent relations between nouns in texts. Accordingly, the degrees of each node will correlate with the noun's frequency. As in all cases of word frequencies, Zipf's Law applies here too (see section 5.3.4). Therefore, the degree distributions are similar across all texts. Since most nouns appear only once (or very infrequently) in texts, they will seldom co-occur with other nouns. This results in sparse networks, with low density measures.

In chapters 6 and 7 we identified several characteristics of Stein's literary style, such as a limited vocabulary and many repetitions. As a result, there are relatively more co-occurrences of nouns in her novel, and this is reflected in the relatively high density of the network associated with this novel. Another manifestation of this is the large weight of some of the edges of the network. The "heaviest" edge which is not a loop connects the noun 'kind' to the noun 'way' and weighs 109, indicating the number of sentences in which these two nouns co-occur. This weight is three times greater than the heaviest edge in Stein's autobiography and the heaviest edge in Wharton's autobiography, and seven times larger than the heaviest edge in Wharton's novel.

Centrality measures allow us to zoom in and examine the content rather than the general quantifiable style of the texts. In both autobiographies, words indicating time, such as 'day', 'year', and 'time', stand out, as would be expected of this specific genre. The novels, however, are not similar. Wharton's novel is characterized by words that pertain to the senses: 'eye', 'sense', and 'hand'. Accordingly, it can be assumed that the centrality of the word moment also stems from the focus on sensuality, a theme that Wharton often re-visits in her writing (Erlich 1992). In contrast, the key nouns in Stein's novel are very general: 'thing', 'girl', 'home', 'woman', and 'man'. This conforms with the modernistic style attributed to Stein.

Of course, there is no validity to this interpretation without going back to the text and reading the books themselves. The computational process is not the end of the interpretive process, but rather its beginning. Once we identify the key nouns we can return to the original texts, locate the context in which these words were expressed, and then interpret their centrality.

This example of distant reading concludes our journey. It exemplifies many of the features that characterize meaningful computational explorations in the humanities, as we have experienced throughout the book. This type of study is always a process that must be performed critically, with an awareness of potential biases, and informed evaluation of the outcome. This process is iterative by nature, should always lead back to the source materials for close reading and interpretation, and can be used as a springboard to further investigation.

Glossary

Betweenness Centrality A measure of the extent to which a node lies on the shortest paths between other nodes. It reflects the role of a node as a bridge or broker within the network.

Clique A subset of a network where each node is directly connected to all others.

Closeness Centrality A measure of how close a node is to all other nodes in the network, calculated by the average distance from the node to every other node.

Clustering Coefficient A measure of how connected a node's neighbors are to each other, indicating the degree of clustering within a network.

Degree Centrality The simplest measure of centrality, based on the number of edges (connections) a node has. Higher degree centrality indicates greater importance or influence in the network.

Degree Distribution A way of understanding a network's structure by looking at the frequency of various degrees among nodes.

Density The ratio of actual edges in a network to the total possible edges between nodes, used to assess how connected a network is.

Eigenvector Centrality A measure of a node's importance based on the centrality of its neighbors. It assigns higher scores to nodes connected to other well-connected nodes.

Lemmatization A linguistic process that reduces words to their base or root form.

Modular Decomposition A method of dividing a network into smaller, tightly-knit communities or modules. These communities are groups of nodes with more connections within the group than with nodes outside the group. Modular decomposition helps reveal underlying structures and patterns in large networks.

PageRank A centrality measure originally developed for ranking web pages. It calculates the likelihood that a random walker on the network will land on a particular node, prioritizing nodes connected to other important nodes.

Scale-free Distribution A type of distribution where most nodes have few connections, but a few nodes have many, following a power-law pattern.

References

Alberich, Ricardo, Joe Miro-Julia, and Francesc Rosselló. 2002. "Marvel Universe Looks Almost Like a Real Social Network". https://arxiv.org/abs/cond-mat/0202174.

Bavelas, Alex. 1950. "Communication Patterns in Task-Oriented Groups". *The Journal of the Acoustical Society of America* 22(6): 725–730. https://doi.org/10.1121/1.1906679

BBC. 2021. "Suez Canal: Ships Stuck in 'Traffic Jam' as Salvage Efforts Continue". March 27. www.bbc.com/news/world-middle-east-56538653.

Erlich, Gloria C. 1992. *The Sexual Education of Edith Wharton.* University of California Press.

Freeman, Linton. 1977. "A Set of Measures of Centrality Based on Betweenness". *Sociometry* 40(1): 35–41.

He, Kaiming, Xiangyu Zhang, Shaoqing Ren, and Jian Sun. 2016. "Deep Residual Learning for Image Recognition". *Proceedings of the IEEE Conference on Computer Vision and Pattern Recognition*, 770–778. Piscataway, NJ: IEEE. https://doi.org/10.1109/CVPR.2016.90

Jackson, Cornell. 2017. "Using Social Network to Reveal Unseen Relationships in Medieval Scotland". *Digital Scholarship in the Humanities* 32(2): 336–343. https://doi.org/10.1093/llc/fqv070

Jenkins, David. 2001. "A Network Analysis of Inca Roads, Administrative Centers, and Storage Facilities". *Ethnohistory* 48(4): 655–687. https://doi.org/10.1215/00141801-48-4-655

Jockers, Matthew. 2012 "Computing and Visualizing the 19th-Century Literary Genome". *Digital Humanities Conference 2012*, 242–244, University of Hamburg.

Karinthy, Frigyes. 1929. *Minden Másképpen Van Ötvenkét Vasárnap.* Athenaeum.

Kilcoyne, Martin Joseph. 1961. "The Political Influence of Rasputin," PhD diss. University of Washington.

Landau, Edmund. 1895. "Zur relativen Wertbemessung der Turnierresultate". *Deutsches Wochenschach* 11: 366–369.

Milgram, Stanley. 1967. "The Small World Problem". *Psychology Today* 2: 60–67.

Moretti, Franco. 2011. "Network Theory, Plot Analysis". *New Left Review* 68. https://newlef treview.org/issues/ii68/articles/franco-moretti-network-theory-plot-analysis.

Page, Lawrence, Sergey Brin, Rajeev Motwani and Terry Winograd. 1999. *The PageRank Citation Ranking: Bringing Order to the Web*. Technical Report 1999-66. Stanford InfoLab.

Raykoff, Ivan, and Robert Deam Tobin (Eds.). 2007. *A Song for Europe: Popular Music and Politics in the Eurovision Song Contest*. Ashgate.

Shapira, Assaf. 2020. "How Do I Know if I am The Center of the Network, or a Network Overview of The Game of Thrones". February 17. www.snapod.net/post/%D7%94%D7%A8%D7%97%D7%91%D7%95%D7%AA-%D7%9C%D7%A4%D7%A8%D7%A7-4-%D7%9E%D7%93%D7%93%D7%99-%D7%9E%D7%A8%D7%9B%D7%96%D7%99%D7%95%D7%AA-%D7%90%D7%95-%D7%9E%D7%91%D7%98-%D7%A8%D7%A9%D7%AA%D7%99-%D7%A2%D7%9C-%D7%9E%D7%A9%D7%97%D7%A7%D7%99-%D7%94%D7%9B%D7%A1

Skorinkin, Daniil. 2017. "Extracting Character Networks to Explore Literary Plot Dynamics". *Proceedings of Dialogue: Conference on Linguistic Computing* (*Komp'juternaja Linguistics i Intellektual'nye Tehnologii*), Issue 16, Vol. 1257–1270.

Stinsky, Daniel. 2021. "'Integration, Nobody Knows What It Means': European Cooperation and the United Nations Economic Commission for Europe (UNECE), 1947–56". In *European Integration beyond Brussels: Unity in East and West Europe since 1945*, edited by Matthew Broad and Suvi Kansikas. Palgrave Macmillan.

Trilcke, Peer, Frank Fischer, Mathias Göbel and Dario Kampkaspar. 2015. "Comedy vs. Tragedy: Network Values by Genre. Network Analysis of Dramatic Texts". July 31. https://dlina.github.io/Network-Values-by-Genre/.https://dlina.github.io/Network-Val ues-by-Genre/

Watson, Joey, and Amruta Slee. 2019. "Eurovision Was Meant to be Apolitical. It's Been Anything But". *ABC Radio National*, May 15. www.abc.net.au/news/2019-05-16/eurovis ion-political-and-diplomatic-history-of-song-contest/11109018.

Young, Nicholas Maurice, Binod Sundararajan, Mary Liz Stewart and Paul Stewart. 2009. "Even Superheroes Need a Network: Harriet Tubman and The Rise of Insurgency in the New York State Underground Railroad". *Du Bois Review: Social Science Research on Race* 6(2): 397–429. https://doi.org/10.1017/S1742058X09990208

Solutions to selected questions

1. Creating a network based on two CSV files:

```
import networkx as nx
import csv
nodes = []
node_names = []
edges = []
```

```
# read the nodes file and create a list of node names
nodecsv = open("euroroad_nodes.csv","r")
nodereader = csv.reader(nodecsv) # Read the csv

for n in list(nodereader)[1:]:
    nodes.append(n)
nodecsv.close()

for n in nodes:
    node_names.append(n[0])

# read the edges file and create a list of edge tuples
edgecsv = open("euroroad_edges.csv","r")
nodereader = csv.reader(edgecsv)

for row in list(nodereader)[1:]:
    t = tuple(row)
    edges.append(t)

edgecsv.close()

# create the network
G = nx.Graph()
G.add_nodes_from(node_names)
G.add_edges_from(edges)
```

2. The shortest route between two cities:

```
city1 = input("City 1? ")
city2 = input("City 2? ")
path  = nx.shortest_path(G,  source=city1,  target=
city2)
path_len = len(path)
print("The shortest path between " + city1 + " and
" + city2 + " is ", path)
print("Its length is ", path_len)
```

3. The correct answers are a, b, and f, since the degrees of all of these stations is 4.
4. Nodes (2) and (3) have a closeness centrality of 1, while (4) has a centrality of 0.75.

5. The correct answers are e, f, g, and h.
6. Connectedness and connected components
    a. The number of components in the E-road network:

```
print("The   network   has   ",   nx.number_connected_
components(G), " components.")
```

    b. Components and nodes

```
import matplotlib.pyplot as plt
for i in nx.connected_components(G):
    subG = G.subgraph(i)
    print(subG.number_of_nodes(), " nodes:")
    nx.draw(subG, with_labels=True)
    plt.show()
```

7. Degree centrality
    a. The 10 most central nodes in terms of degree centrality:

```
deg_list = []
for node in G.nodes():
    deg = G.degree(node)
    deg_list.append((deg,node))

deg_list.sort(reverse=True)

for n in deg_list[:10]:
    print(n)
```

8. A subgraph of the largest component in the E-road network

```
components = nx.connected_components(G)
largest_component = max(components, key=len)
subG = G.subgraph(largest_component)
```

9. Closeness centrality
   a. The five most and five least central cities in terms of closeness

```
close_dict = nx.closeness_centrality(subG)
close_list = []
for k, v in close_dict.items():
    close_list.append((v, k))
close_list.sort(reverse=True)
print("5 nodes with greatest closeness score:\n",
close_list[:5])
print("5 nodes with smallest closeness score:\n",
close_list[-5:])
```

10. Betweenness centrality
    a. The five most and five least central cities in terms of betweenness

```
between_dict = nx.betweenness_centrality(subG)
between_list = []
for k, v in between_dict.items():
    between_list.append((v, k))
between_list.sort(reverse=True)
print("5 nodes with greatest betweenness score:\n",
between_list[:5])
print("5 nodes with smallest betweenness score:\n",
between_list[-5:])
```

11. Eigenvector centrality
    a. The five most and five least central cities in terms of the eigenvector
       measure

```
eigenvector_dict = nx.eigenvector_centrality(subG,
max_iter=500)
eigenvector_list = []
for k, v in eigenvector_dict.items():
    eigenvector_list.append((v, k))
eigenvector_list.sort(reverse=True)
print("5 nodes with greatest eigenvector score: \n",
eigenvector_list[:5])
print("5 nodes with smallest eigenvector score: \n",
eigenvector_list[-5:])
```

Note that by default, NetworkX is limited to 100 iterations for computing eigen-vector centrality. However, depending on the network structure and on the algorithm, this number of iterations may not be enough. When this is the case, the program returns an error message. To override this limitation, we can add as a second argument to eigenvector_centrality() the definition max_iter=500, which increases the maximal number of iterations.

12. The Eurovision network
   a. Creating the Eurovision network

```
import networkx as nx
import csv
import community.community_louvain as cm

edges=[]

# read the edges file and create a list of edge tuples
edgecsv = open("eurovision_edges.csv","r")
edgereader = csv.reader(edgecsv)
for row in list(edgereader)[1:]:
    edges.append((row[0], row[1], float(row[2])))
edgecsv.close()

# create the network
G = nx.Graph()
G.add_weighted_edges_from(edges)
```

   b. Dividing the network into communities:

```
partition = cm.best_partition(G,resolution = 0.6)
communities = {}
for node in partition:
    val = partition.get(node)
    if val in communities:
        communities[val].append(node)
    else:
        communities[val] = [node]
```

   c. Printing the communities:

```
for i in communities:
    print(i,communities[i])
```

14. Building the noun network:
    a. Cleaning the text

```
cleantext = rawtext.replace("\n"," ")
cleantext = cleantext.replace(" .",".")
```

   b. Creating the noun co-occurrence network

```
import nltk
nltk.download("punkt")
nltk.download("averaged_perceptron_tagger")
nltk.download("wordnet")
from nltk.stem import WordNetLemmatizer
wnl = WordNetLemmatizer()
import matplotlib.pyplot as plt
import networkx as nx

tokens = []
tagged = []
nountext = []
sentences = nltk.sent_tokenize(cleantext)
#==== STEP 1: Extracting nouns ======
# iterate over all sentences
for sent in sentences:
    nounsent = []
    tokens = nltk.word_tokenize(sent)
    tagged = nltk.pos_tag(tokens)
# iterate over all words in a sentence
    for word in tagged:
        if word[1] == "NN" or word[1] == "NNS":
            nounsent.append(wnl.lemmatize   (word[0].
            lower()))
    #   append the list of nouns (nounsent) to the
    list of lists (nountext)
    if len(nounsent) > 0:
        nountext.append(nounsent)

#==== STEP 2: Calculating co-occurrences of nouns in
sentences ======
d = {}
```

```
#  iterate over all sentences in nountext
for sent in nountext:
#  iterate over all nouns in a sentence
    for i in range(len(sent)):
        token = sent[i]
        next_token = sent[i+1:]
#  iterate over next words and create/update keys
        for t in next_token:
            key = tuple (sorted([t, token]))
            if d.get(key):
                d[key] += 1
            else:
                d[key] = 1
#==== Step 3: Creating a list of edges ======
noun_noun_edges = []
for key in d:
    noun_noun_edges.append((key[0], key[1],d[key]))

#==== Step 4: Creating the network ======
G = nx.Graph()
G.add_weighted_edges_from(noun_noun_edges)
```

15. Analyzing the networks
    a. The 10 "heaviest" edges

```
edge_weight_list = []
for u,v,z in noun_noun_edges:
    edge_weight_list.append((z,u,v))
edge_weight_list.sort(reverse=True)
for n in edge_weight_list[:10]:
    print(n)
```

   b. The top 10 nodes by degree centrality

```
deg_list = []
for node in G.nodes():
    deg_list.append((G.degree(node),node))
```

```
deg_list.sort(reverse=True)
for n in deg_list[:10]:
    print(n)
```

c. Connectedness & components

```
if nx.is_connected(G):
    print("The network is connected.")
else:
    print("The network has ", nx.number_connected_
    components(G), " components.")
components = nx.connected_components(G)
largest_component = max(components, key=len)
subG = G.subgraph(largest_component)
print("The largest component has : " + str(subG.
number_of_nodes()) + " nodes and " + str(subG.
number_of_edges()) + " edges.")
```

d. The top 10 nodes by closeness centrality

```
close_dict = {}
close_dict = nx.closeness_centrality(subG)
close_list = []
for i in close_dict:
    close_list.append((close_dict[i], i))
close_list.sort(reverse=True)
for n in close_list[:10]:
    print(n)
```

e. The density of the network

```
print("The density of the network is ", nx.density(G))
```

f. Degree distribution

```
degrees = []
for n in G.nodes():
    degrees.append(G.degree(n))
plt.hist(degrees)
plt.show()
```

Note

1 The data used in this chapter are taken from http://konect.cc/networks/subelj_euroroad, and are updated for 2017.

Index